T0211504

SpringerBriefs in Applied Sciences and Technology

Series editor

Janusz Kacprzyk, Systems Research Institute, Polish Academy of Sciences, Warsaw, Poland

SpringerBriefs present concise summaries of cutting-edge research and practical applications across a wide spectrum of fields. Featuring compact volumes of 50 to 125 pages, the series covers a range of content from professional to academic. Typical publications can be:

A timely report of state-of-the art methods
An introduction to or a manual for the application of mathematical or computer techniques
A bridge between new research results, as published in journal articles
A snapshot of a hot or emerging topic
An in-depth case study
A presentation of core concepts that students must understand in order to make independent contributions

SpringerBriefs are characterized by fast, global electronic dissemination, standard publishing contracts, standardized manuscript preparation and formatting guidelines, and expedited production schedules.

On the one hand, **SpringerBriefs in Applied Sciences and Technology** are devoted to the publication of fundamentals and applications within the different classical engineering disciplines as well as in interdisciplinary fields that recently emerged between these areas. On the other hand, as the boundary separating fundamental research and applied technology is more and more dissolving, this series is particularly open to trans-disciplinary topics between fundamental science and engineering.

Indexed by EI-Compendex and Springerlink

More information about this series at http://www.springer.com/series/8884

Juraj Ružbarský · Anton Panda

Plasma and Thermal Spraying

Juraj Ružbarský
Faculty of Manufacturing Technologies
Technical University of Košice
Prešov
Slovakia

Anton Panda
Faculty of Manufacturing Technologies
Technical University of Košice
Prešov
Slovakia

This monograph has been supported by the projects KEGA 027TUKE-4/2014 and VEGA 1/0381/15.

ISSN 2191-530X   ISSN 2191-5318   (electronic)
SpringerBriefs in Applied Sciences and Technology
ISBN 978-3-319-46272-1   ISBN 978-3-319-46273-8   (eBook)
DOI 10.1007/978-3-319-46273-8

Library of Congress Control Number: 2016952880

Printed on acid-free paper

This Springer imprint is published by Springer Nature
The registered company is Springer International Publishing AG
The registered company address is: Gewerbestrasse 11, 6330 Cham, Switzerland

Contents

Introduction

In case of all technologies of thermic application, the methods and possibilities of improvement in diverse properties of applied coatings have been constantly sought for, which appear logical as it represents one of the ways of enhancement of the application of usually rather expensive material of high quality. Moreover, new knowledge related to the sphere of applied coatings is gained that can serve for improvement in surface properties of basic material. In the practice of mechanical engineering, the film coatings are applied onto metal material by being dipped into liquid metal, by thermal and plasma spraying or by other technologies.

Thermic application of films is used for the surface treatment of new products, for the renovation of components, for the purpose of service life prolongation and for the purpose of increase in wear resistance as well as in thermal insulation and in corrosion resistance in different environment types. Technology of thermic spraying allows broad scale of materials to be sprayed. Those are materials on the basis of iron and of non-ferrous metals and their alloys, and ceramic materials and their alloys. The materials must conform to strict requirements yet should be affordable.

The technology of plasma spraying can be regarded as the improved development level of flame metallization. The metalizing process was invented and patented in 1910 by Max Ulrich Schoop in Switzerland. At the same time Schoop was the inventor of the method of arc spraying. The arc spraying was suppressed by flame metalizing which started to develop at the beginning of the thirties in the USA. Its usage expanded by spraying of anticorrosive coatings of zinc and aluminium and included was also renovation of machinery components by spraying of steel and bronze. An effort is made to apply powder metallization the development of which commenced in the fifties in connection with fast progress of powder materials, thus opening new possibilities for the application of metallization. Then, apart from hardsurfacing materials, the oxides of aluminium and zirconium become interesting.

After the Second World War, the methods of arc spraying developed rapidly in Central Europe. From the start, the flame and arc metallization were limited by low-temperature oxygen and acetylene flame or electric arc which means that, for instance, the material with melting point of 2700 °C could not be sprayed. During this period rather developed was thermal spraying by flame. In spraying with auxiliary material in the form of wire, the particles are pulled by the jet of compressed air out of the slag bath so the grain size depends on spraying parameters. In spraying with auxiliary material in the form of powder, the size of particles is determined by their grain composition. When impinging on the basic material, the molten particle is deformed under the influence of its high kinetic energy. Elastic deformation occurs in the particle during the impingement, and only by the effect of impulse pressure, it is deformed on the basic material surface.

In case of thermal spraying by electric arc between two fed wires out of which one represents the anode and the other is the cathode an electric arc is formed. The electric arc heat melts the auxiliary material—wires in the area of the anode and of the cathode spot. The compressed air flowing out of the nozzle at high velocity drags the molten metal particles and casts them onto the basic material.

The development of further method of plasma spraying followed. The method brought another advantage in the form of spraying in the environment of inert gasses considerably decreasing oxidation of molten particles both during the flow through the aeriform atmosphere and landing onto the basic material. The temperatures reached by standard plasma torch are substantially higher contrary to the melting temperatures of all known materials. In majority of spraying processes the optimal temperature range is from 6500 up to 11000 °C.

Plasma spraying allows application of broad assortment of coatings—ceramic coatings, refractory metals, combinations of alloys and plastics, etc.—with minimal thermal influence upon the basic material. Coating material in form of powder is fed by the flow of gas, argon, nitrogen, hydrogen, helium, or by their combination into the plasma jet generated by electric arc burning. In plasma produced by the torch, the ions are connected with the electrons releasing abundance of energy which heats the gas up to over 6000 °C, causing fast expansion of gas. Plasma flame melts the powder, and gas expansion shoots the molten drops out onto the coated surface at velocity of even 300 m.s^{-1} in case of which the drops solidify rapidly.

In reference to successful usage, especially to service life of the applied coating, the attention focuses on adhesion of the coating acquired as described afore in dependence of its thickness on basic surface. The presented thesis analyses the issue of adhesion as well.

Chapter 1
Plasma Jet

1.1 Characteristics of Plasma Spraying

In 1928 an American physicist Langmuir described plasma as the gas state which besides neutral atoms and molecules contains positive and negative particles, ions and electrons. Plasma does not follow the conventional laws of thermodynamics therefore was regarded by a number of scientists as the fourth state of matter differing from solid, liquid, and gas phases. Plasma state often occurs in the nature. In the Sun eruption the ejected plasma clouds consist of nuclei of hydrogen atoms and of electrons. They arrive at our Earth at velocity of 1500 km h^{-1} causing the northern lights, magnetic storms, and ionospheric disturbances. Several natural and laboratory plasmas exist that are characterized by number of electrons in cm^3, by temperature of electrons (K) and by induction of magnetic field (T) which can be categorized as follows:

Plasma type	Number of electrons in 1 cm^3	Temperature of electrons (K)	Induction of magnetic field (T)
Interstellar matter	$1–10^3$	10^4	10^{-9}
Ionosphere	$10^3–10^5$	2×10^3	5×10^{-5}
Solar corona	10^8	10^6	$10–10^{-6}$
Electric arc	$10^{16}–10^{18}$	10^4	$10^{-4}–10^{-7}$
Electromagnetic pulses	$10^{15}–10^{18}$	5×10^4	10^{-9}
Thermonuclear discharge	10^{16}	$10^6–10^8$	$10^{-7}–10^{-9}$
Matter inside the Sun	$10^{22}–10^{25}$	10^7	–

J. Ružbarský and A. Panda, *Plasma and Thermal Spraying*,
SpringerBriefs in Applied Sciences and Technology,
DOI 10.1007/978-3-319-46273-8_1

Plasma differs from common gases by the following properties:

- in plasma the chemically homogeneous gas changes into a mixture of diverse types of particles—molecules, atoms, positive and negative ions, electrons and photons,
- apart from elastic collisions of molecules applied are the inelastic ones which causes the changes of quantic states of molecules (dissociation, ionization),
- plasma contains charged particles—electrons and ions which can affect the gas through electromagnetic field; gases are conductive and send or receive energy to/from magnetic field. The mixture as the whole is electrically quasi neutral.

The electric arc represents the most widespread source of high-pressure plasma. In plasma spraying the fed inert gas (or water) is heated by the heat produced by electric arc and dissociates from molar to atomic state. For instance:

$$N_2 + V_D \rightarrow 2N \tag{1.1}$$

$$H_2 + V_D \rightarrow 2H \tag{1.2}$$

with V_D—dissociation energy

For diverse gases there exist diverse energy types and ionizations; for hydrogen 4.477 eV, or 13.5 eV, for nitrogen 9.76 eV, or 14.5 eV, for oxygen 5.08 eV, or 13.6 eV, for CO_2 16.56 eV is energy of dissociation, and for argon 15.7 eV is energy of ionization (1 eV = 1.602 × 10^{-19} J). To maintain a stable arc in such state inevitable is to increase the intensity of electric field (increases with growth of amount of fed gas) by which the temperature in the middle of a column (temperature in plasma axis increases along with increasing current and voltage). In case of these higher temperatures the gas atoms are ionized as follows:

$$N + V_i = N^+ + e \tag{1.3}$$

$$H + V_i = H^+ + e \tag{1.4}$$

$$Ar + V_i = Ar^+ + e \tag{1.5}$$

with V_i—ionization energy given for diverse gases.

A number of processes start (diffusion, transmission—conduction of heat, chemical reactions) during formation of plasma jet in the stable electric arch. Atoms from hotter plasma areas, in which the dissociation degree is higher along with higher number of atoms (of dissociated molecules), diffuse to colder areas with undissociated molecules (for instance, close to the walls of cooled tube). In these areas the atoms are recombined and release recombining heat with recombination energy equal to energy of dissociation of molecules and to energy of ionization of atoms. At the same temperature enthalpy of the diatomic gases, for instance of hydrogen, is higher contrary to monoatomic gases such as argon. Molecules again diffuse to hotter areas and consume certain amount of heat. Plasma is thus formed in the environment with dissociation, ionization, and recombination processes. Therefore plasma is referred to as a specific state of matter with concurrent occurrence of ions, electrons, photons, atoms, and molecules of gas. Its characteristic feature is quasi neutrality, i.e. concentration of positive and negative particles in plasma is identical and the final charge of space equals in fact to zero. Thermal conductivity of such gas composes of contact and diffusion heat (in dissociated and ionized gas). Two methods of formation of plasma jet are distinguished as follows:

(a) by transferred arc (Fig. 1.1),
(b) by non-transferred arc (Fig. 1.2).

In plasma formation by transferred arc the electric arc occurs between the wolfram (graphitic) cathode and auxiliary material (wire) as the anode. The arc is stabilized by sulphur of ionizing gas flowing out of the torch nozzle which pulls the formed plasma jet. This method is applied in welding and cutting of metal by plasma.

In case of the second plasma formation the arc burns between the wolfram cathode and the wall of water cooled nozzle included in anode. The electric arc heats up the fed working gas by high pressure to high temperature which results in dissociation, ionization, and recombination in plasma formation. Likewise the plasma arc, the plasma itself flows out of the nozzle at high velocity. The system is

Fig. 1.1 Scheme of plasma torch with transferred arc

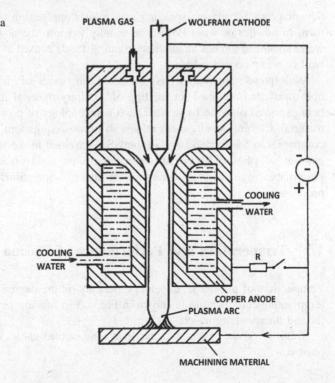

Fig. 1.2 Scheme of plasma torch with non-transferred arc

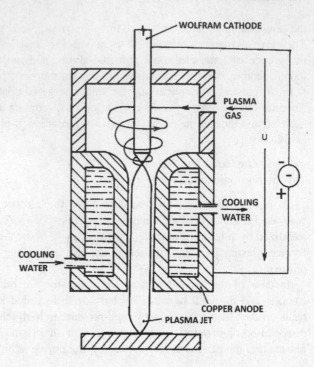

advantageous for plasma spraying as it allows application of auxiliary material in form of powder or wire. Powder is usually fed into the nozzle orifice. In case of water stabilized torches an auxiliary rotating anode is used to prevent burning of the wall of water cooled nozzle.

Widespread application is observed in case of a plasma torch with non-transferred arc used for melting of auxiliary material in form of powder and being sprayed onto the basic surface. It is technology of powder metal and ceramic material. Current development related to plasma equipment production by several companies in Switzerland and in the USA provided in the world market sufficient amount of plasma spray automatic machines and robots controlled by programmable regulation in assurance of diverse operations in serial large-scale production.

1.2 Temperature and Performance of Plasma Jet

Temperature of plasma jet depends especially on the degree of ionization. Typical temperature classification is shown in Fig. 1.3 in relation to laminar argon plasma jet and turbulent argon jet.

Table 1.1 shows mean and maximal temperature values of argon and nitrogen plasma.

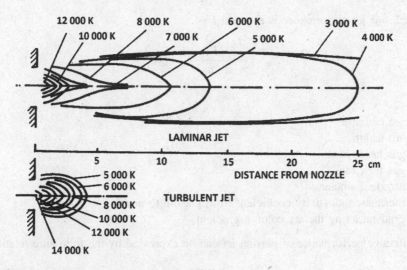

Fig. 1.3 Temperature classification for plasma jet

Table 1.1 Mean and maximal temperature values of argon and nitrogen plasma

Performance (kW)	Temperature (K)	Flow rate					
		15 N l min^{-1}		30 N l min^{-1}		60 N l min^{-1}	
		Argon	Nitrogen	Argon	Nitrogen	Argon	Nitrogen
5	T_{max}	11 400	–	11 750	–	12 000	–
	T_{str}	8 850	–	5 600	–	3 000	–
10	T_{max}	–	10 900	12 700	11 040	13 000	10 540
	T_{str}	10 750	7 100	9 600	6 250	6 600	5 550
15	T_{max}	–	12 370	14 500	11 990	14 400	12 180
	T_{str}	12 000	7 550	11 000	6 700	8 750	5 950
20	T_{max}	–	13 830	–	14 790	–	13 100
	T_{str}	–	8 250	–	7 100	–	6 200

Nl stands for 1 l of gas in standard state, i.e. 0 °C and 101.3 kPa. Mean values of exit temperature of plasma as well as of performance of plasma jet can be expressed by the following theoretical relations:

$$T_q = \frac{\bar{E} \cdot I}{\pi \cdot d \cdot a_k} \left[(1 - \exp) - \frac{\pi \cdot d \cdot a_k \cdot L}{C_p \cdot \overline{Q}} \right] \tag{1.6}$$

If the relation in square brackets is denoted by letter K the plasma jet temperature is expressed as

$$T_q = \frac{\bar{E} \cdot I}{\pi \cdot d \cdot a_k} \cdot K \tag{1.7}$$

and plasma jet performance is expressed as

$$N_q = \frac{\bar{E} \cdot I}{\pi \cdot d \cdot a_k} \cdot C_p \cdot \bar{Q} \cdot K \tag{1.8}$$

with

I arc current,
L arc length,
C_p gas latent heat,
Q gas flow rate,
d nozzle diameter,
a_k thermal conductivity coefficient from plasma to nozzle,
E gradient along the arc column gradient.

Effective performance of plasma jet can be expressed by the following relation

$$N_{ef} = 0.24U \cdot I \cdot \eta \tag{1.9}$$

with

U arc voltage,
I arc current,
η efficiency of electric energy use for heating of gases.

1.3   Exhaust Velocity

Dependence of velocity prior to and after gas heating is obtained from the equation of constant flow rate as follows

$$\rho_o \cdot v_o = \rho \cdot v \tag{1.10}$$

and from the following equation of state

$$\frac{p_o}{\rho_o} \cdot T_o = \frac{p}{\rho} \cdot T \tag{1.11}$$

with

ρ_o and ρ gas density at inlet and outlet of nozzle,
p_o and p pressure at inlet and outlet of nozzle or of cold and heated gas,
T_o and T temperature of cold and heated gas in K determined according to enthalpy,
v_o and v current velocity at inlet and outlet of nozzle.

Provided that in gas heating its pressure remains constant the following is applicable

$$\frac{v_o}{T_o} = \frac{v}{T} \tag{1.12}$$

out of which

$$v = v_o \cdot \frac{T}{T_o} \tag{1.13}$$

Such dependence provides the values which correspond to values obtained in the experiment. The most frequently used experimental method is based upon measurement of flexibility of current out of which velocity is determined according to impulse sentence as follows

$$F = m \cdot v \tag{1.14}$$

with

F flexibility of current,
m weight flow rate of current.

Velocity of plasma jet can be calculated also in dependence on performance of plasma jet and its properties and in dependence on nozzle diameter as per the following relation

$$v = A \cdot \frac{Q_o}{d^2} \cdot \frac{T}{M} \left(\text{m s}^{-1} \right) \tag{1.15}$$

with

A constant,
Q_o volumetric flow rate of gas ($\text{m}^3 \text{ s}^{-1}$),
T gas temperature (K),
d nozzle diameter (m),
M molecular weight of gas.

Velocities of plasma jet are proportional to flow rate of gas and reciprocally proportional to square of nozzle diameter. The graph shows dependence of velocity of plasma jet on arc current, on flow rate of plasma gas, and on distance from the nozzle (Fig. 1.4).

Plasma jet velocity is influenced by enlargement of flow rate of carrying gas of powder which apart from hydrostatic influence decreases the temperature and velocity of plasma. Due to the structures of modern plasma torches the maximal velocities are reached excessing the velocity of sound. Such high velocities of plasma jet along with its high temperature assure high velocity of thermal transmission towards auxiliary material intended for melting-down and spraying and

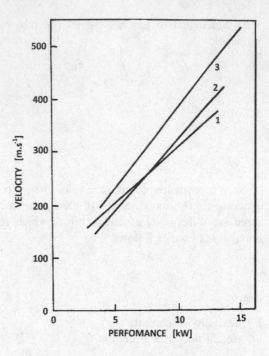

Fig. 1.4 Dependence of plasma jet velocity on the arc current, on the flow rate of gas, and on the distance from the nozzle

impart the high energy to molten particles which positively influences the quality of coating formed during impingements of flowing molten particles onto the basic backplate.

1.4 Plasma Gases

Argon, hydrogen, nitrogen, and water are used in our conditions, and in the USA even helium is applied. Plasma gases are classified into two groups—the monoatomic ones and the diatomic ones. The first group includes argon and helium and the second group contains nitrogen and hydrogen. Basic physical and chemical characteristics of the used plasma gases are presented in Table 1.2.

Gaseous environment in the plasma torch forms the plasma, protects the electrodes against oxidation and cools down the electrodes. Plasma gas is selected in dependence on required temperature and velocity of plasma jet and inertness to sprayed and basic material.

Table 1.2 Physical and chemical characteristics of plasma gases

Characteristics	Dimension	Argon	Helium	Nitrogen	Hydrogen
Relative gram-atomic weight		39.944	4.002	28.016	2.0156
Density at 0 °C and at 101.32 kPa	$(kg\ m^{-3})$	1.783	0.1785	1.2505	0.0898
Specific thermal capacity C_p at 20 °C	$(kJ\ kg^{-1}\ K^{-1})$	0.511	5.233	1.046	14.268
Coefficient of thermal conductivity at 0 °C	$(W\ m^{-1}\ K^{-1})$	0.01633	0.14363	0.02386	0.17543
Potential of ionization	(V)	15.7	24.05	14.5	13.5
Potential of one-stage ionization	(V)	27.5	54.1	29.4	–
Potential of two-stage ionization	(K)	14,000	20,000	7300	5100
Temperature	(V)	40	47	60	62
Arc voltage	(kW)	48	50	65	120
Power supplied to arc coefficient of energy usage for gas heating	(%)	40	48	60	80

1.4.1   Dissociation, Ionization and Enthalpy of Plasma Gases

Difference between the process of plasma formation in case of monoatomic and diatomic gas rests in the fact that ionization of atoms of diatomic gas begins after dissociation of its molecules—hydrogen at temperature of 5000 K and nitrogen at 8500 K. Further significant difference between monoatomic and diatomic gases is diverse enthalpy and temperature of plasma formed out of those gases. Figure 1.5 shows nitrogen with enthalpy 5 times higher contrary to argon at temperature of 8000 K as the energy obtained by monoatomic gases in plasma column is determined by specific heat and ionization energy whereas in case of diatomic gases abundance of energy is obtained through dissociation of molecules to atoms.

1.4.1.1   Basic Characteristic Properties of Plasma Gases

Nitrogen

Dissociation begins at 5000 K, and at 9000 K the dissociation degree reaches 95 %. Ionization begins at 8000 K, and at 15,000 K the ionization degree reaches 50 %, at 20,000 K the ionization degree reaches 95 %, and at 22,000 K the ionization degree reaches 98 %. From 7000 K nitrogen plasma contains higher amount of heat per unit of mass contrary to other gases. Higher enthalpy and longer plasma jet of nitrogen plasma allow simpler application of refractory material unlike

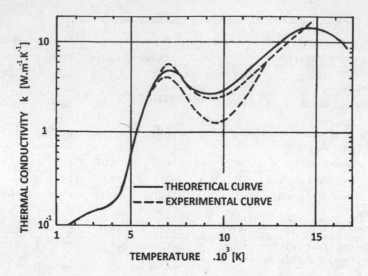

Fig. 1.5 Dependence of thermal conductivity of nitrogen plasma on temperature

in case of argon plasma. For the same reason the longer distance between a torch and a sprayed object must be observed contrary to argon plasma. The used nitrogen cannot contain admixtures of oxygen as poisonous nitrogen oxide is formed and electrodes oxidize.

Hydrogen

Dissociation occurs from 2000 K up to 6000 K. Atomic hydrogen is completely ionized at 25,000 K. Dissociation of hydrogen requires less energy unlike dissociation of nitrogen. Moreover, hydrogen enthalpy is lower contrary to the nitrogen one. Hydrogen possesses high thermal conductivity and hydrogen plasma needs the highest arc voltage and maximal supplied power out of all gases. Despite the aforementioned the temperature of hydrogen plasma is the lowest one contrary to other plasma types. The plasma torch must be thoroughly tight since presence of oxygen in hydrogen can result in formation of the explosive mixture.

Argon and Helium

Ionization of argon begins at 9000 K and is completed at 22,000 K. Enthalpy of these monoatomic gases is substantially lower contrary to the diatomic ones. However, transmission to plasma state is simpler, they produce stable electric arc, and they require lower operating voltage. Their temperature reaches the highest values in comparison with other plasma types. Argon and especially helium assure high inertness of environment from the point of view of influence upon material. Argon and helium do not dissociate, their plasma jet is shorter and contracted and shines brightly. Such jet allows localized application of sprayed coating which means considerably higher effectiveness of application. Helium possesses high thermal conductivity. The price is high as helium ranks among rare gases.

Table 1.3 Influence of composition of plasma mixtures upon performance of plasma jet

	Plasma formed from gas mixture	
	Ar + H$_2$	Ar + H$_2$
Content of argon (%)	66	66
Overall flow rate (l min^{-1})	35	35
Operating current (A)	110	110
Operating voltage (V)	40	65
Power (kVA)	4.4	7.14

To increase enthalpy and velocity of plasma jet the mixtures of gases in diverse dose rates are used. Mixture of argon and nitrogen as of the heaviest gases is suitable for increasing of kinetic energy of the jet and of the sprayed and molten material. To increase the velocity of plasma jet nitrogen along with 3 even 10 % of hydrogen must be used. To assure inertness of environment the most frequent is use of mixture of argon and hydrogen. Influence of composition of plasma gases upon performance of plasma arc can be documented in case of two mixture regimes of plasma-forming gas (Table 1.3).

Enthalpy and temperature of gases can be regulated within wide range by the change of supplied electric input, of flow rate, and of composition of plasma gases in the plasma torch.

1.5   Thermal Conductivity of Plasma

Thermal conductivity is determined by the following quantities:

k_c conventional conductivity through micromovement of atoms and ions,
k_E conventional conductivity through movement of electrons,
k_D effect of diffusion of associated atom pairs (ambipolar diffusion),
k_j^+ effect of diffusion of ions and electrons,
k_j^{++} effect of separation into doubly ionized atoms and electrons.

Summary effect of these quantities at diverse temperatures is expressed by the curves in Fig. 1.5.

Rather high is thermal conductivity of hydrogen and of helium which is only two times lower contrary to the thermal conductivity of copper. Therefore plasma mixtures containing high percentage of the gases may considerably wear the electrodes out, especially the anodes, since standard cooling is not sufficient for them.

1.6   Electric Conductivity of Plasma

Electric conductivity of plasma with high temperature depends on presence of charged particles, on mobility of electrons being 100 times higher contrary to mobility of ions. In case of the first approximation it can be assumed that electricity

Fig. 1.6 Electric
conductivity of plasma of Ar,
He, N_2 and H_2 in dependence
on temperature

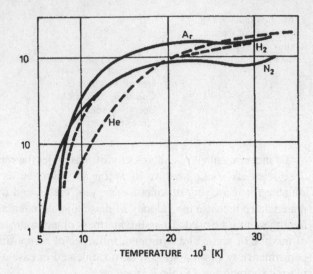

conducting is caused by electrons and the following may be applicable for electric
conductivity

$$\lambda = e \cdot n_e \cdot b_e \tag{16}$$

with

e elementary electric charge,
n_e electron density,
b_e electron mobility.

Figure 1.6 shows electric conductivity of nitrogen, argon, helium, and hydrogen
plasma in graph.

Electric conductivity of nitrogen and argon plasma do not change over
20,000 K and become equal to electric conductivity of metal conductors.

Chapter 2
Formation of Plasma Sprayed Coating

2.1 Powder Transport, Flow Trajectory and Velocity of Sprayed Particles

Suitable powder for plasma spraying should possess good flow properties, should easily flow in plasma jet; the powder particles should be of spherical or similar shape and should dispose of identical dimensions defined within the range of the lower limit. In powder spraying with particles of diverse dimension the fine fractions are overheated or evaporated and coarse-grained particles dot not get heated. Important role is played by the orifice of powder feed into plasma jet. The highest effectiveness of powder heating is achieved during its feeding into the centre of the arc burning between cathode and anode in the torch. Such alternative is complicated as to structure. Therefore in majority of modern equipment the powder fed by the flow of carrying gas enters the plasma jet radially from the external part of the nozzle—from the anode.

The powder particle is forced perpendicularly onto the plasma jet axis and copies the trajectory dependent on diverse factors. Particle jet moving to the basic material is of conical shape with angle γ inclining to the plasma jet axis by angle α (Fig. 2.1).

The angle depends on input velocity of particles flowing into the plasma jet and on their size as well as on plasma jet velocity and on classification of velocity in cross section of the plasma jet in the point of particle inflow. Minimal values of angle α assure favourable conditions for spraying as under given circumstances the number of grains decreases. The grains move along the external and colder plasma jet trajectory.

The shape of particle trajectory is determining for the spraying effectiveness. Trajectory of particles depends on a number of factors. The most significant ones include geometry of plasma torch and of powder injector, type and flow rate of plasma and carrying gas, powder characteristics, arc performance, stability of plasma jet, etc. Velocity of molten powder particles is directly proportional to velocity of plasma jet and is regulated by the same parameters as the velocity of plasma jet.

© The Author(s) 2017
J. Ružbarský and A. Panda, *Plasma and Thermal Spraying*,
SpringerBriefs in Applied Sciences and Technology,
DOI 10.1007/978-3-319-46273-8_2

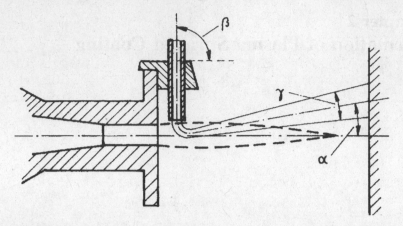

Fig. 2.1 Inclination of the jet of molten particles in the plasma jet

Diverse optic and kinematographic methods allow determination of particle velocity in the experiment. The experiments proved that bigger particles possessed lower velocity when flowing out of the plasma jet yet losing is more complicated. Vice versa, smaller particles are easily accelerated and decelerated by the ambient atmosphere. Detected was also considerable variance of particles which can be explained by radial distribution of velocities in plasma jet and by variability of the individual grains. The relations between velocity and diameter are detected experimentally. In spraying of fine-grained material higher velocities are reached. With increasing distance from the nozzle deceleration of the lighter particles is more frequent contrary to the heavier ones. The particles are accelerated just by certain distance from the nozzle. Both optimal spraying distance, at which the velocity reaches maximal value, and the mean value of this velocity depend on dynamic properties of plasma jet. Their regulation is possible by the torch performance, by the nozzle geometry, by the flow rate of plasma gases, etc.

Calculating method of determination of particle velocity in the plasma jet can stem from the following relation:

$$v = v_{pl}\left[1 - \exp\left(-\frac{2A \cdot l}{v_{pl}}\right)\right] \qquad (2.1)$$

with

v particle velocity,
v_{pl} plasma jet velocity,
l particle trajectory,
A $4.5\ \eta.r^2.\rho$ is constant,
η environment viscosity,

r particle diameter,
ρ particle material density.

Velocity calculated as per the relation mentioned afore must be regarded as the approximate one.

2.2 Mutual Influence of Molten Material with Plasma Jet and Ambient Atmosphere

Under the influence of thermal and kinetic effects of plasma jet the fed powder is molten down and accelerated. Mechanism of transmission of heat from the plasma jet to powder is rather complicated phenomenon and has not been clarified up to present. In case of the first approximation an assumption can be made that energy transmitted to powder q is proportional to the temperature, to the length of thermally active zone of plasma jet, and to the coefficient of transmission of heat. The energy is indirectly proportional to the velocity of plasma jet as per the following relation:

$$q = \frac{T \cdot \alpha \cdot l}{v} \qquad (2.2)$$

With

q effective output of heat,
T plasma jet temperature,
α coefficient of transmission of heat,
l length of active thermal zone of plasma jet,
v plasma jet velocity.

According to the relation mentioned afore the cause of faulty heating can rest in short time of powder residence in plasma ($T - 1/v = 10^{-14}$ up to 10^{-2} s), insufficient intensity of heat exchange between plasma and powder or difficulties with forcing of powder into high-temperature part of jet.

In examination of powder heating inevitable is to take into consideration the Bi criterion which is determined by the relation between plasma transmission of heat λ_p and particle material λ_l:

$$Bi = \frac{\lambda_p}{\lambda_l} \qquad (2.3)$$

The expression $Bi < 1$ (Ar, He, N_2) is applicable for plasma with low thermal conductivity. For plasma with good conductivity (H_2) in heating of oxides and other material with low coefficient λ_l applicable is $Bi > 1$ meaning that for plasmas with the coefficient of $Bi < 1$ the transmission of heat is limited by the processes in the jet in plasma of other group in heating of non-conducting powders by transmission of heat into the depth of particle.

If powder overheating in case of which the temperature in the centre of the particle reaches 0.9 multiple of material melting temperature is considered to be sufficient obtained can be the following expression for determining of maximal cross section of particles which can heat up in plasma:

$$d_{max} = \sqrt{\frac{4 \cdot a \cdot \tau}{0,3}} \qquad (2.4)$$

With

a coefficient of thermal conductivity $\left(a = \frac{\lambda}{c \cdot \rho} \right)$,

τ time of particle residence in plasma,

λ coefficient of thermal conductivity,

c specific heat,

ρ density.

Derived was the relation for minimal critical diameter of particles which can still reach the sprayed surface:

$$d_{max} = \sqrt{\frac{18 \cdot \mu \cdot k_{krit} \cdot l}{\gamma_l \cdot v_\infty}} \qquad (2.5)$$

with

μ plasma viscosity,

γ_l density of particle material,

k_{krit} critical value of the Stokes number:

$$k_{krit} = \frac{\gamma_l \cdot d \cdot v_\infty}{18 \cdot v \cdot l} \qquad (2.6)$$

with

d particle diameter,

v_∞ plasma jet velocity,

v viscosity coefficient,

l torch distance from the backplate.

Dependence of appropriate adjustment of powder and of coating quality on grain composition of powder was proved in the experiment by other authors with determining of particle temperature in plasma in dependence on grain composition of powder and on distance from the nozzle (Fig. 2.2). According to the graph powder grains of WC-Co 40 μm are evaporated and grains of 150 μm remain intact.

Fig. 2.2 Temperature of
particles of WC-Co in
dependence on grain
composition of powder and
on distance of nozzle in argon
plasma jet

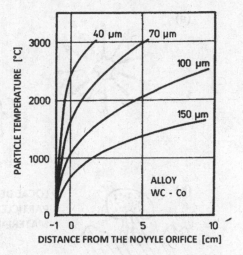

Pursuant to these facts 'only the powders with strictly limited dimensions (diameters) offer reliable results.

Along the trajectory in plasma the particles interact with gases of ambient atmosphere and with plasma as all material types demonstrate high chemical activity at melting temperature or at temperatures reaching the melting one. A range of mechanisms of reciprocal interaction of sprayed particles with gases include gas adsorption, chemical mutual reaction and formation of oxide layers and other linkages on the surface of particles, gas dissolution in liquid metal of particles, diffusion processes and mechanical mixing of the products of mutual interaction in particle volume, etc.

2.3   Impingement of Flowing Particles onto the Backplate

During impingement of flowing molten particles onto the backplate their kinetic energy changes into thermal and deformation energy. The processes in the zone of impingement can be theoretically explained by the theory of hydrodynamic impingement of liquid particles onto the hard surface. Pursuant to the theses regarding the sphere in question it is apparent that during impingement of particles kinetic energy causes their deformation and generation of substantial pressures— the dynamic and the impulse ones. At the first moment in the point of contact with the surface the liquid particle is deformed. In the time interval ranging from 10^{-10} up to 10^{-6} a thin layer of diameter D is formed approximating the particle diameter d. Further on the particles are deformed evenly (Fig. 2.3).

Formation of flat coating is connected with impulse pressure p_1 as a result of movement of pressure waves spreading within the particle from the moment of

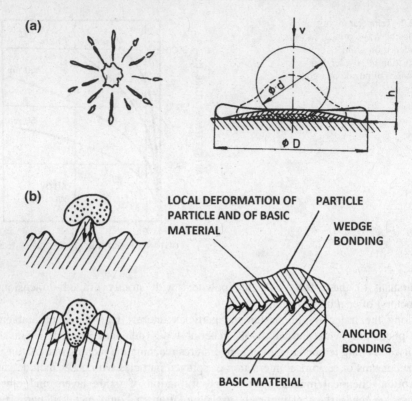

Fig. 2.3 Scheme of deformation of particles on the backplate **a** smooth backplate, **b** rough backplate

impinging onto the surface. Maximal impulse pressure can be expressed by the following expression:

$$p_1 = \frac{\mu}{2} \cdot \rho_{kv} \cdot c_{zv} \cdot v \tag{2.7}$$

with

μ coefficient of particle consistency,
ρ_{kv} liquid density,
v particle velocity,
c_{zv} sound velocity in the liquid.

If $c_{zv} = (2 \text{ up to } 5) \cdot 10^3 \text{ m s}^{-1}$ is applicable for molten metals and if $\mu = 0.5$, $\rho_{kv} = 40 \text{ g cm}^{-3}$, $v = 100 \text{ m s}^{-1}$ is accepted for spraying conditions, so consequently $p_1 = 0.98$ Gpa.

Dynamic pressure of current flowing evenly is calculated according to Bernoulli's equation. If assumed is pressure p_d acting upon the part of surface of a backplate dimensionally proximate to particle diameter and if the backplate is considered to be incompressible and the particle is regarded as ideal liquid the simplest case of impingement is achieved with dynamic pressure

$$p_d = \rho_{kv} \cdot v^2 \tag{2.8}$$

Duration of pressure effect upon particle axes

$$\tau = (d - h) \cdot v \tag{2.9}$$

with h—height of particle solidified during time τ.

Under real conditions of spraying the velocity of particles reaches 100 even 300 m s^{-1} and the values of dynamic pressure can range from 49 up to 98 MPa in the course of effect duration 10^{-5} up to 10^{-7}.

During impingement onto basic backplate the particles which under influence of surface stress take spherical shape and deform substantially deliver their kinetic and thermal energy and solidify in a shape of lamellae or lens with thickness ranging from 5 up to 15 µm. By laying up the lamellae the layer with the characteristic lamellar structure is formed (Fig. 2.4).

Molten drops falling onto, for instance, smooth glass surface burst fast and spread along the backplate area (Fig. 2.3a). The rough surface is prevented from quick spreading of drops by rapid cooling and surface unevenness which causes capturing of the solidifying drops (Fig. 2.3b).

Particles being individually cooled and crystalized generate heterogeneous structure. Apart from boundaries typical for compact materials the sprayed coating includes the boundaries between the deformed particles and the individual coatings (Fig. 2.4). The conditions of formation of boundaries between the coatings differ from the conditions of formation of boundaries among particles especially in the

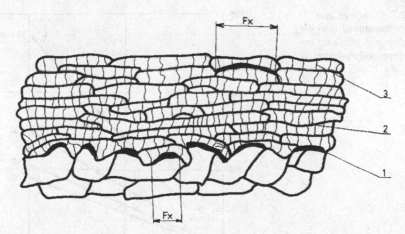

Fig. 2.4 Scheme of film structure. *1* interface between the film and the basis, *2* interface between the individual coatings, *3* interface between particles, F_x—diameter of the surface section onto which a particle was welded

course of duration of contact with the atmosphere. In spraying of the individual coatings the gas adsorption and evaporation of the finest fractions of sprayed material occur on preceding coating. The conditions related to deformation and to bonding with the basic backplate or with preceding coating are not met by all particles during the impingement. The particles reflect from or settle down on the surface as unmolten powder and defects and pores are formed along with material loses and proportion of non-deformed particles is increased. All of the phenomena and specificities of the coating formation in spraying cause that, for instance, hardness of metal coatings reaches 10 even 20 % of hardness of similar compact material.

2.4 Thermic Influence upon the Backplate and Sprayed Coating

Transmitted heat consists of the heat transmitted by plasma jet Q_p and by particle Q_m. These values were thoroughly examined by Kudinov according to whom total amount of heat transmitted onto the backplate per unit of time Q increases linearly with the input supplied into the torch and decreases rapidly with the extending distance of torch from the backplate (Fig. 2.5).

Relation of total amount of heat to supplied input P_e, i.e. to efficiency, is shown in Fig. 2.5 in dependence on spraying distance L. In such case the following is applicable

Fig. 2.5 Change of heat amount transmitted onto the backplate in dependence on torch input and spraying distance L

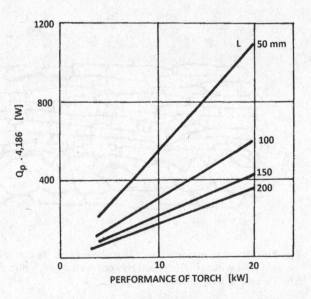

$$\eta = \frac{Q}{P_e} \qquad (2.10)$$

$$\eta_m = \frac{Q_m}{P_e} \qquad (2.11)$$

and

$$\eta = \eta_p + \eta_m \qquad (2.12)$$

According to Fig. 2.6 the heat transmitted by plasma jet decreases considerably faster in dependence on extending of distance contrary to heat transmitted from particle material. To prevent formation of slag bath the sufficient spraying distance must be selected. Adequate spraying distance is 100 even 150 mm as shown in Fig. 2.6.

The temperature of contact between the deforming molten-down particle in impingement and hard cold backplate is expressed by the following relation

$$T_k = \frac{T_m \cdot K_E}{K_L + \phi(d)} \qquad (2.13)$$

$$K_E = \frac{\lambda_p}{\lambda_n} \cdot \sqrt{\frac{a_n}{a_p}} \qquad (2.14)$$

$$K_L = \frac{c_p \cdot T_m}{1.77 \cdot L} \qquad (2.15)$$

with

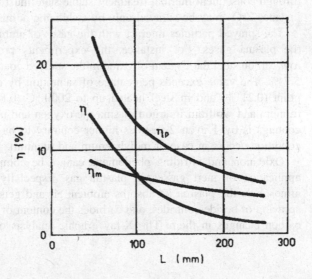

Fig. 2.6 Efficiency of transmission of heat in dependence on spraying distance

K_E coefficient characterizing thermal activity of particle with regards to the backplate,

K_L criterion for evaluation of latent heat L of particle material,

λ_p coefficient of thermal conductivity of particle,

λ_n coefficient of thermal conductivity of backplate,

a_p, a_n coefficients of thermal conductivity of particle and backplate,

c_p specific heat of particle,

T_m melting temperature of particle,

$\emptyset(d)$ function of probability integral.

2.5 Physical and Chemical Changes of Material in Spraying

Subliming materials with boiling point approximating melting point can be sprayed only with certain difficulties. MgO dissociates and disintegrating products are volatile. The disintegration could be prevented by spraying of such materials in argon plasma enriched by 20 % oxygen. Several oxides, for instance ZrO_2, may partially get reduced during spraying in clear and air-free plasma. In spraying $ZrSiO_4$ dissociates to ZrO_2 and SiO_2.

Polymorphic materials are exposed to allotropic changes and fast cooling after spraying can cause their metastabilization. Under high temperature heated γ–Al_2O_3 changes into cubic α–Al_2O_3 which due to rapid cooling acquires unstable structure of passing state referred to as δ–Al_2O_3. By adding nitrogen the δ—crystal lattice can be stabilized. In case of materials the properties of which are changed by phase modifications the danger of the sprayed coating disturbance occurs and thus through consequent thermal treatment stable state must be restored. Therefore, for instance ZrO_2 may be sprayed only by stabilizing admixture of—5 % CaO.

The sprayed particles interact with the gases of ambient environment and with the plasma gases. For instance, the experiments proved that in spraying of low-carbon steel the content of oxide in the sprayed coating reaches from 2.5 up to 5.7 %. The value exceeds percentage of saturation by oxygen in steel at melting point (0.21 %) and in steel heated up to 2000 °C (0.87 %). In spraying of aluminium and wolfram in argon plasma the oxygen and nitrogen content in sprayed coatings is by 1 even 2 degrees higher contrary to basic material. The identical condition occurs in case of molybdenum and titanium.

Oxidation and nitriding phenomena cannot be completely prevented even by application of inert gasses in plasma gas. Especially in spraying in aeriform atmosphere the plasma jet absorbs ambient air and gets mixed rapidly with it. In spraying of borides, nitrides, and carbides the content of boron, of nitrogen, and of carbon changes in them. The X-ray graphic analysis of the sprayed coatings of carbides of zircon, of niobium, and of tantalum in atmosphere with oxygen content of <11 % detected just a single phase—carbide in homogeneous sphere. By increasing of oxygen content the two phases were detected—in case of zircon carbide it was zircon oxide and carbide oxide, in case of niobium carbide those were oxides of Nb_2O_5 and NbO_2. In case of wolfram carbide the traces of metal wolfram were identified at high temperature as per reaction $2WC = W_2C + C$. The coatings of zircon carbides and of niobium carbides sprayed in environment with 4.5 % oxygen demonstrated a single-phase structure.

Chapter 3
Basic Properties of Plasma Coatings

Plasma sprayed films acquire properties differing from those of basic compact materials: oxygen and nitrogen content increases and density and plasticity decrease.

3.1 Structure

Particles molten down in plasma jet acquire spherical shape by surface stress. After impingement onto the basic material the molten particles are stacked up in a pancake or a lamella pattern unless the film is formed. Kinetic energy of flowing particles changes into heat during impingement.

Plasma spray coatings possess completely different structure contrary to corresponding compact materials. Due to different structure the mechanical and physical properties of sprayed coatings are dissimilar.

Internal structure of the film is not homogeneous and consists of reciprocally adhered multi-grains. The films formed from the molten particles are typical for lamellar structure of coatings. The structure of coatings formed in molten and solid phase only slightly differs from the structure of compact materials.

Superior structure with lower porosity and with better bonding between the individual coatings and basis is achieved by application in the inert gas or the vacuum chamber. By annealing to recrystallization temperature of metals, the lamellar structure changes into the grained one. Contrary to plasma spray coatings in atmosphere, the structure contains no inclusions and oxides.

3.2 Density and Porosity

Significant indicators of the structure include porosity. The more liquid the drops of sprayed metal are and the higher the velocity of flowing towards the basis is, the denser the film structure is. Plasma spraying torches by means of which high flowing velocity of plasma jet is reached can help achieve the coatings with rather low porosity. Contrary to this, fragile and hard spray coating materials dispose of high porosity.

Density of plasma films ranges from 85 up to 93 % of theoretical density of the identical liquid material.

Micro-porosity relates to liberation of oxygen, nitrogen, and hydrogen by decline in solubility with temperature decrease.

3.3 Bonding, Internal Stress, Coating Thickness

The sprayed coatings adhere to the surface of basic material especially by interaction of mechanical anchoring, valence and Van der Waals forces. The sprayed films are more fragile contrary to corresponding compact materials and powders of hard materials provide also tensile films. Therefore the metal films are rather tensile and the ceramic ones are fragile. In case of ceramic films the metallic bonding with the basis surface represented mainly by metals is not possible. Thus adhesion of ceramic films is lower as well. In sprayed metal coatings applied on metals the local diffusion processes were proved. For all types of sprayed coatings it is applicable that their adhesion to the basis decreases with increasing thickness of coatings.

The values of adhesion of the sprayed coatings flow below the values of tensile strength of respective materials with the values of 100 MPa regarded as the good ones and of 50 MPa regarded as the sufficient ones. Ceramic coatings with thickness of 0.1 mm reach the adhesion values of 30 up to 40 MPa, the coatings with thickness of up to 0.3 mm reach the adhesion values of approximately 5 up to 10 MPa and the levels with thickness of up to 0.5 mm reach the adhesion value of approximately 4 up to 7 MPa. Decreasing of the coating adhesion with the increasing thickness is caused by internal stress occurring with different melting temperatures and with different coefficients of expansibility as well as the film formed in tiny coatings. Formation of cracks in the sprayed films is related to stresses caused by rapid cooling as well as with the coating structure. Due to economic and technical reasons the coatings should be applied as thin as possible.

Formation of complicated field of internal stresses depends on uneven distribution of the applied material and uneven heating of the component by the local character of the plasma torch effect. At the same time apparent is influence of particularities of the shape and dimensions of the sprayed component itself. Internal stresses are calculated according to distribution of temperature fields in the component until the moment corresponding to zero stresses in the film, i.e. until the moment of its formation. Internal stresses on the film surface are generated by a sample cooling to the ambient temperature. In case of tough gripping of the sample a simplified formulation for thermal stress calculation can be taken into consideration.

$$\sigma_p = \alpha_p \cdot T_p \cdot E_p \tag{3.1}$$

with

α_p coefficient of linear expansibility,
E_p elastic modulus of pressure,
T_p temperature.

Index p refers to quantities related to the film surface. Usually, the thickness of the basis exceeds the one of the film therefore internal stresses are not extensive.

When the coefficient of thermal expansibility of the sprayed material equals to or exceeds coefficient of thermal expansibility of basis material the internal tensile stresses occur. The exception to the rule is the case of Mo spraying onto steel and aluminium. At present the elaborated engineering methods for internal stress calculation are not at disposal.

To decrease internal stresses the basis can be preheated or cooled in spraying. In spraying of easily oxidized materials the basis is recommended to be cooled. Occurring internal stresses can be well compensated by the intermediate coatings, e.g. of NiAl, etc.

3.4   Strength, Hardness, Deformability

Strength and adhesion of the film is affected by its thickness. The stress analysis proves that film thickness increases, and the stress accumulates along with decrease of strength as shown in Fig. 3.1.

In such event σ_x characterizes maximal strength of bonding possible to be achieved among materials by spraying. In case of majority of materials high stresses of the 1st and of the 2nd degree act in the contact zone which considerably weakens the bonding of particles.

Area of spots S_x that can be expressed by the coefficient as follows

$$\beta = \frac{S_x}{S} \tag{3.2}$$

In such event both β and σ_x characterize the strength of particle bonding of the first tiny coating with the basis. Spraying of other tiny coating causes formation of internal stress inside the film σ_o which accumulates with the increase of number of tiny coatings or of thickness of film σ. Therefore in practice decreasing strength of bonding $\beta\sigma_x - \sigma_o$ occurs. As the increase of number of tiny coatings causes

Fig. 3.1 Scheme of change
of coating strength in
dependence on its thickness

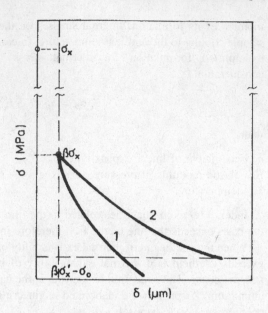

accumulation of internal stresses that may exceed the strength of bonding $\beta\sigma_x-\sigma_o$ the film gets separated from the basis spontaneously (Fig. 3.1, curve 1). In consequence of the film elasticity the strength of bonding is stabilized (Fig. 3.1, curve 2) on certain level (interrupted line). The stresses inside the film become relaxed and do not exceed the values of $\beta\sigma_x$.

Through the change of spraying parameters the film strength can be regulated in the area of curves 1 and 2. In dependence on relation in the area between the strength of the film bonding with the basis and with the particles an adhesive or a cohesive failure may occur inside the film. The weakest point is usually the zone of bonding among the film particles adhering to the basis therefore in case of failure part of the film remains on the backplate surface. Decrease of film strength in this zone is explained by negative influence of high thermal conductivity of compact massive basis upon the contact temperature of the particles of the second, of the third and of other tiny coatings in the proximity of the backplate. With the increase of the film thickness the thermal conductivity decreases, the temperature in the point of bonding and the strength of the film increase. The film thickness typical for occurrence of such phenomena ranges from 0.5 up to 1 mm.

Achieved are the strengths of sprayed coatings and only fractions of values of analogic compact materials. In case of molybdenum bending strength of 260 MPA was measured and in case of wolfram the value reached 340 MPA. Due to thermal processing the values increased 1.5 even 2 times.

In measurement of hardness inevitable is to take into consideration the fact that that especially with thin coatings the thickness must be at least 10 multiple of penetration depth and measurements according to Vickers require the thickness to be 2 multiple of penetration diagonal. With hardness of HRV 1000 and loading of

3 N the coating must be 0.05 mm thick and with hardness of HRV 500 the thickness must reach at least 0.07 mm. Under certain circumstances the hardness of thin coating should be measured with loading ranging from 0.05 up to 0.5 N. To achieve comparable values the hardness with loading of 0.1 up to 0.5 N is measured.

The hardness values of WC coatings range from HRV 700 up to 1350. Dense coatings of Al_2O_3 reach the hardness of approximately HRV 1000 and less dense coatings reach the hardness of HRV 700. Hardness of Ti coating sprayed in the air by means of nitrogen plasma is HRV 700 and in spraying in the inert chamber by Ar plasma the coating hardness is just HRV 270. Substantial difference related to hardness of coatings of reactive metals may occur in spraying by plasma jet generated by diverse gases. The particles of WC or of Al_2O_3 in the sprayed coatings are considerably harder contrary to HRV ranging from 2000 up to 2800.

Deformability of the sprayed coatings depends on type of material and in case of fragile coatings even on coating thickness. Adhesion of coating and properties of basic material are significant as well.

3.5   Thermal and Electric Conductivity

Heat in the plasma sprayed films can be transmitted as follows:

(a) by electrons of metal particles of which the film consists even in sections of bonding with the formed strong metallic bonding (λe),
(b) by lattice or photon thermal conductivity in particles which is significant for non-metal materials and according to chemical bonds among particles (λ_f),
(c) by photons emitted in film pores (λ_l) if the film is heated to high temperature,
(d) by thermal conductivity of gas enclosed in the pores of the film (λ_m).

Since the heat is transmitted slowly as described in (c), the summary coefficient of thermal conductivity of films

$$\lambda = \lambda_e + \lambda_f + \lambda_l + \lambda_m \tag{3.3}$$

is lower contrary to the one of respective compact materials. Comparison of thermal conductivity coefficients proved that thermal conductivity of the films is almost by one degree lower unlike the thermal conductivity of the identical materials in contact form. Apart from this, the thermal conductivity of films disposes of complicated dependence on temperature.

The determined effective value of the coefficient of thermal conductivity of plasma sprayed coating of powder $ZrSiO_4$ ranges $\lambda_{(T-25\,°C)} = 0.211-0.037$ W m^{-1} K^{-1}. In case of several applications of plasma sprayed coatings required are low or minimal values of thermal and electric conductivity. These properties are considerably influenced by spraying conditions. For thermally and electrically insulating coating the respective technical parameters require the most functional thickness of the sprayed coating. For instance, densely sprayed coating of Al_2O_3 disposes of dielectric strength of 5000 V per millimetre of thickness. For coatings of Mo and Cu sprayed in the air the values of electric conductivity are approximately of 1/10 of the values which are applicable for these metals in compact state. In spraying in the inert chamber substantially higher values can be reached.

Ceramic films are used for thermal-insulating purposes at high temperatures. Especially frequent is application of Al_2O_3, of Cr_2O_3, of TiO_2, of pure or stabilized ZrO_2 with 5 % of CaO, etc. Fine Al_2O_3 is used for electrically insulating films to obtain dense coatings.

Chapter 4
Adhesion of Plasma Sprayed Coatings to Basic Backplate

4.1 Types of Bonding Forces

The existing scientific experience take into consideration the influence of the following factor in formation of adhesion mechanism:

(a) mechanical anchoring of liquid particles of sprayed metal into basic material,
(b) application of forces of physical interaction of the type by Van der Waals,
(c) chemical interaction directed towards microwelding of particles, i.e. to occurrence of strong chemical bonds of covalent or metal type,
(d) metallurgical processes occurring in the immediate proximity of formed microwelds.

Two possible variants of such type of bonding is are distinguished (Fig. 4.1):

(a) wedge-shaped attachment if dimension of micro-unevenness of basic material is smaller contrary to the one at the peak of a projection and if bonding is formed just by friction forces,
(b) anchor attachment if micro-unevenness is distinguished at the basis. To generate attachment of anchor type with the utmost strength of bonding between the coating and the backplate inevitable is to assure maximal fluidity and velocity of applied particles.

A number of observers reckon that bonding of the coating with basic material is of complex character and is conditioned by mechanical attachment as well as by formation of chemical bonds between the coating and backplate. If in spraying of ceramic coatings onto metal the mechanical bonding types prevail, in spraying of metal onto metal the role of metal bond increases.

© The Author(s) 2017
J. Ružbarský and A. Panda, *Plasma and Thermal Spraying*,
SpringerBriefs in Applied Sciences and Technology,
DOI 10.1007/978-3-319-46273-8_4

31

Fig. 4.1 Types of bonding between the coating and the backplate

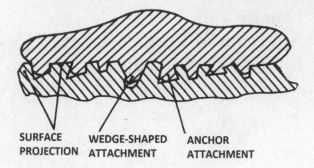

SURFACE   WEDGE-SHAPED   ANCHOR
PROJECTION   ATTACHMENT   ATTACHMENT

4.2   Theoretical Aspects of Optimal Adhesion Formation

According to several authors the process of formation of chemical bonds among atoms of adhering surfaces is characterized by two phases as follows:

(a) approximation of adhering surfaces to reach the distance inevitable for inter-action of atoms,
(b) quantic processes of electron interaction.

It can be assumed that following the impingement onto the sprayed surface the particle cools down and becomes dispersed at a rapid rate with concurrent lifting of the front of solidification from the point of contact. Such condition leads to the particle maintaining certain pressure during solidification which approximates and compresses interacting phases. The pressure reaches the values ranging from 8 up to 15 MPa by means of which the particle deforms plastically along with upper coating of the backplate. The most significant outcome of the analysis is the knowledge referring to maintaining of the temperature higher than the backplate temperature for certain time. Inevitable is to assess whether the period of time is sufficient for formation of bonding between the coating and the backplate. Consequently, the following relation is applicable

$$t_a = \frac{1}{v} \ln \left(1 - \frac{N}{N_0} \right) \exp \frac{E_a}{k \cdot T_k} \tag{4.1}$$

with

t_a length of the process of reaction actuation,
v frequency of inherent oscillations of atoms,
$\frac{N}{N_0}$ mutual relation of reacting atoms and all atoms occurring on the surface,
E_a activated energy of the process of bonding formation,
k empiric constant

Figure 4.2 shows the principle of application of quantity t_a. Curve 1 charac-terizes the dependence of reaction time on temperature. Decreasing of temperature prolongs the time inevitable of course of reaction. Change of temperature over time

Fig. 4.2 Dependence of
length of actuation process on
temperature

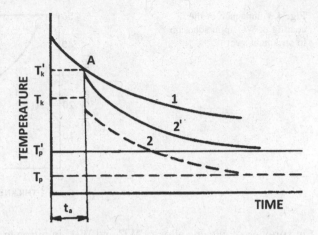

is characterized by the curves 2 and 2'. Completion of the total crystallization of the
particle is followed by temperature decrease from T_k—contact temperature—to
T_p—backplate temperature. The course of reaction requires intersection or at least
meeting of the curves 1 and 2 which means that time of residing of the particle at
constant temperature is sufficient for formation of chemical bonds with the
backplate.

Just the curve 2 can be regulated as the shape of the curve 1 is determined by
physical and chemical properties.

To calculate the adhesion the following relation in general form is applicable:

$$F_{sp} = \frac{A_m}{r} - f(\tau) \qquad (4.2)$$

with

A_m adhesive work of the coating,
r radius of action of adhesive forces,
$f(\tau)$ combined function of dimension of stress in the coating

With high values of $f(\tau)$ the force F_{sp} can be decreased to zero or it can become
negative. In such case the coating detaches by action of internal shearing forces.

4.3 Principal Factors Influencing Adhesion of Coating to Basic Surface

4.3.1 Thickness of Coating

Thickness of coating negatively influences the strength of bonding between the
coating and backplate which is proved by the results of experimental measurements

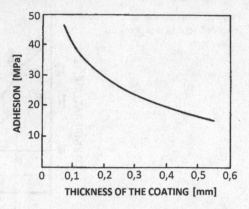

Fig. 4.3 Influence of the coating of WC upon adhesion to structural steel

in spraying of eutectic alloy of WC and W_2C in nitrogen plasma onto the structural steel (Fig. 4.3).

In quantitative expression of decrease of bonding force between the sprayed coating and the backplate with increase of thickness of coating it can be assumed that stress in the coating spreads in the same way as in circular plate freely bearing against the plane. Consequently, the equation intended for metal coatings can be compiled as follows:

$$F = -K \cdot \delta^2 \tag{4.3}$$

with

F bonding force inevitable for detaching of the coating,
K coefficient of proportionality,
δ thickness of the coating

Figure 4.4 shows stress distribution in the system of coating—backplate in case of thin and thick coating.

With increase of thickness the maximal bend angle decreases which the coating can bear without a crack during its deformation by means of flexural deformation. Such technological test represents indirect indicator of strength of bonding between the coating and the backplate.

4.3.2   Pretreatment of Basic Material Surface

Pretreatment should provide the surface coatings with positive geometric and energetic properties for formation of strong bonding between the spray coating material and the basic material from the point of view of mechanical anchoring as well as from the point of view of microwelds occurrence. In fact, the surface treatment includes degreasing, roughing, and cleaning the area by chemical

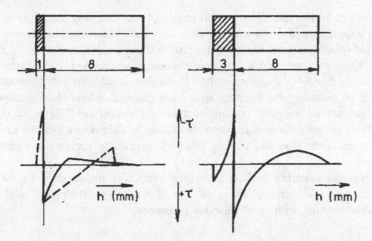

Fig. 4.4 Stress distribution in the system of coating—backplate in case of thin and thick coating

solutions, by abrasive cleaning or by treatment through application of intermediate coating or by preheating.

In cleaning the grains of the abrasive are applied onto the material surface by compressed air jet or by centrifugal force. Irregularities occurring on the surface of basic material are significant from the point of view of mechanical anchoring and for the enlargement of contact surface between the spray coating materials and backplate. The peaks become the active centres of formation of strong bonding as heat dissipation at those points in inconsiderable and time of constant temperature of contact t_k is prolonged. Dependence of curves of actuating energy and of temperature of contact may intersect and strong bonding is achieved at standard room temperature of the backplate.

Preheating of basic material is needed until strong oxide films are formed on the backplate surface. Preheating is useful up to certain critical temperature the value of which ranges from 100 up to 200 °C for majority of commonly applied materials. Further increase of adhesion on the basis of backplate preheating is possible just in case of working under the conditions of controlled atmosphere.

Recent development of technology of plasma spraying pointed out the possibility of further increase of adhesion by spraying of thin intermediate coatings of so-called self-bonding materials. Those are materials the properties of which assure independently appropriate conditions for formation of microwelds. Thus the centre of the problem is shifted from the boundary of basic material—spraying onto the boundary of intermediate coating—covering coating where the conditions for formation of strong bonding are more favourable with respect to distinctive relief and decreased thermal conductivity of the sprayed intermediate coating. Rather unusual properties of molybdenum offered the first possibility to bond diverse sprayings to the basic material through molybdenum coating. Molybdenum spraying caused partial welding with the backplate even in case of smooth areas of basic material.

Other research in the respective branch improved new bonding material—alloy of NiAl by application of exothermic reaction between aluminium and nickel. When nickel clad spherical aluminium particles pass through the plasma jet and get heated to 1400 °C the exothermic reaction occurs in case of which the particle temperature reaches even 3185 °C and thus rather favourable conditions are generated for formation of metallurgical bonding with basic material. Material of intermediate coating can act as damping coating equalizing different coefficients of thermal expansibility of basic material and sprayed coating by means of which the causes of internal stress formation and coating cracking, especially during its cooling, are minimized.

Self-bonding materials include even other refractory metals (Nb, Ta) or compounds causing exothermic reaction (Ni–Ti). Also alloy Ni 80 Cr 20 is used as the intermediate coating with good adhesive properties.

4.3.3 Parameters of Spraying Process

Apart from the main factors mentioned afore the adhesion of the sprayed coatings is affected by parameters of plasma spraying process, especially by jet, stress, distance between torch and backplate, amount of powder and of plasma, and quantity of carrier gas. All of these working parameters influence the generation of conditions assuring thermal and kinetic characteristics of plasma jet, of transport and of melting down of sprayed powder onto the backplate. Since the characteristics influence the process of adhesion of the sprayed material to the backplate the experiment should be used to determine optimum of these parameters for each type and grain composition of the applied powder material. Essentials for determination of optimal parameters are represented by characteristics of the applied powder material (melting temperature, size and shape of grains, oxidizing properties), by characteristics of basic material, by structural and performance characteristics of plasma spraying equipment.

4.4 Experimental Assessment of Adhesion of Plasma Sprayed Coatings to Basic Backplate

4.4.1 Testing Methods of Adhesion Detection

Pursuant to study of diverse methods of adhesion detection and on the basis of own experience the following methods of adhesion assessment are applied:

(a) Technological tests—spray coating detachment by a chisel, by a file, by a hammer or by forcing a ball into the coating surface,
(b) Metallographic analysis of bonding between the coating and the backplate,

(c) Accurate measurement of force inevitable for detachment of the area unit of the coating from the basic material.

Technological tests represent the most simple control methods which are at present applied in practice. Important is to take into consideration the fact that the tests represent a burden for subjective factors and do not allow numerical expression of adhesion and comparison of results.

Metallographic analysis of bonding between the coating and the backplate allows optic view of interface of the backplate and the coating.

Measurement of force inevitable for detachment of area unit of the coating requires utilization of tensile testing machines. In tensile tests and tangentially in shear test or in combination of both methods—in tensile and shear tests. As in Slovakia the adhesion test methodologies of plasma sprayed coatings have not been standardized yet, the experimental measurements of adhesion applied the methods elaborated in Germany as per DIN 50 160 standard and DIN 50 161 standard along with the apex stone method developed by Matejka.

4.4.2   Adhesion Test According to DIN 50 160 Standard

Adhesion of plasma sprayed coatings is assessed in tensile test. The testing sample shape is shown in Fig. 4.5.

According to the standard the diameter of cylindrical part to the front of which the coating is sprayed reaches 40 mm. Series of experiments proved that inconsiderable deviation of the sprayed area diameter from the given value does not influence the measured values of adhesion. Therefore from the point of view of manufacture simplification and of possibility of sample fixation into tensile testing machine by Instron the samples of differing diameter and shape were used (Fig. 4.6).

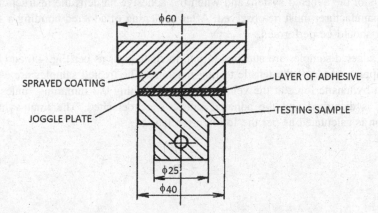

Fig. 4.5 Testing sample for adhesion test according to DIN 50 160 standard

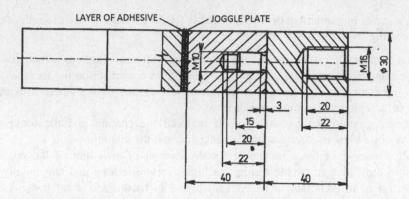

Fig. 4.6 Adjusted shape of a sample for adhesion test according to DIN 50 160 standard

The other front area of such sample disposes of an opening with M 10 series thread into which an intermediate part is screwed bonding the sample with a clamping head of the tensile testing machine.

The smooth front area of the sample is pretreated by abrasive cleaning and chemical degreasing. The plasma sprayed coating with thickness of 1 mm applied onto the area is treated by means of soft lathe-turning to the thickness of 0.8 mm as stipulated by DIN 50 160 standard. Consequently, the samples are adhered to counterparts of the same shape (Fig. 4.6) according to the following procedure:

(a) areas to be adhered are degreased and softly roughened and abrasively cleaned under pressure of 0.3 MPa,

(b) adequate two-component metal adhesive is prepared; in experiments the adhesives Epoxy CHS 1200 and Belzona Standart Metall are used,

(c) during adhesion a thin layer is applied to both areas (of the sample and of the counterpart) and attention should be paid to constant thickness of adhesive and alignment of the sample with the counterpart. When pressure acts upon both parts of the adhered system and when the adhesive hardens the instructions of a manufacturer must be observed. After hardening of adhered bonding a shred test should be performed.

The adhered samples are attached to intermediate parts as per Fig. 4.6 and fixed to a gripping heads of the tensile testing machine. The treated samples are evenly affected by tensile force at the velocity of jaws reaching 0.2 mm.min^{-1} unless the coating is disturbed and the adhered samples are detached. The final value of adhesion is calculated as per the following relation

$$\sigma = \frac{4 \cdot F}{\pi \cdot D^2} \qquad (4.4)$$

With

σ adhesion (tensile strength), (MPa),
F maximal tensile force as set by the tensile testing machine (N),
D diameter of the testing sample with plasma sprayed coating (mm)

To assess the adhesion the mean value σ of three measurements is regarded as the decisive one.

4.4.3   Adhesion Test According to DIN 50 161 Standard

The test measures the adhesion of coatings stressed by shearing forces. The shape and dimensions of the testing sample adjusted in contrast to standard to assure increase of the number of results of a single measurement are shown in Fig. 4.7.

The sample of the specified basic material is made in the shape of pin with several O-rings. The sample surface is treated as per DIN 8 567 Standard and sprayed with coating with thickness ranging from 2 up to 2.5 mm. After the coating is sprayed and cooled the sample shape with four O-rings of the sprayed coating is finished by lathe-turning or by grinding. The height of each O-ring and their diameter are measured with the precision to 0.1 mm.

In the tensile testing machine the sample (Fig. 4.7) is forced into the hardened casing the shape and dimensions of which are shown in Fig. 4.2. Compressive force acts in the sample axis direction and produces shearing force acting upon O-rings of

Fig. 4.7 Shape and dimensions of the adhesion test sample according to DIN 50 161 standard

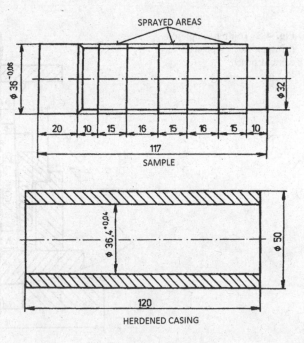

the sprayed coating. The sample is forced into the hardened casing unless the bottom O-ring is cut off. Its removal is followed by stressing of further O-ring.

Adhesion of coating is calculated according to the relation as follows

$$\tau = \frac{F}{\pi \cdot D \cdot h} \tag{4.5}$$

with

F maximal shearing force (N),
h height of O-ring (mm),
D diameter of pin—O-ring (mm)

Thus maximal number of four results of a single test can be achieved.

4.4.4 Adhesion Test by the Apex Stone Method

The test measures adhesion of the sprayed coatings by combination of stressing of the coating by tensile and shearing forces.

The testing sample for such adhesion test type consists of the external part—of the apex stone in which a sliding punch placed. According to the method of assessment of a punch distinguished is the sample of the apex stone type with conical or cylindrical punch. Shape and dimensions are shown in Figs. 4.8 and 4.9.

Fig. 4.8 Testing sample of the apex stone type with conical punch

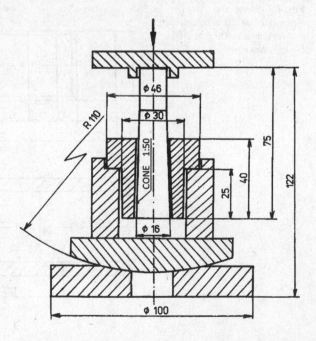

Fig. 4.9 Shape and dimensions of double acting testing sample of the apex stone type with cylindrical punch

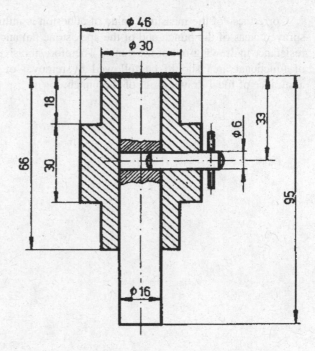

The basis of the test follows from Fig. 4.8: tested coating is sprayed onto the front areas of the apex stone and of the punch pretreated by degreasing and by abrasive cleaning. The front areas are fixed in a single plane during pretreatment and plasma spraying. In case of sample type with the conical punch the fixing is determined by the comicalness of the punch placement. In case of samples with cylindrical punch the punch placement is assured by a safety pin (Fig. 4.9). The sprayed coating with thickness ranging from 1 up to 2 mm is by lathe-turning or by grinding treated to accurate constant value. After release of safety pins the sample is placed into the preparation (Fig. 4.8) and supported in upper extended diameter. In the tensile testing machine the compressive force acts upon the punch in vertical direction. Forcing the punch out of the apex stone the coating during detachment from the basic material is stressed by combination of tensile and shearing forces. The spherical part of preparation (Fig. 4.8) assures perpendicularity of action of compressive force.

Adhesion of coating is calculated according to the relation as follows

$$\sigma = \frac{4 \cdot F}{\pi \cdot (D^2 - d^2)} \tag{4.6}$$

with

F maximal strength detected by testing equipment (N),
D apex stone diameter (mm),
d punch diameter (mm)

Correctness of the measured value of adhesion is influenced by maintaining the sprayed areas of the punch and of the apex stone flat and by removal of additional resistance in drawing-out of the punch. Therefore abrasive cleaning or lathe-turning of functional area should be followed by removal of dust grains, etc. and by assurance of ideal drawing-out of the punch.

Chapter 5
Plasma Spraying Equipment

5.1 Basic Scheme of Plasma Spraying Equipment

Plasma spraying equipment composing of a complex of the individual machines and devices disposing of the following functions: in plasma torch connected by unidirectional current the electric arc is ignited between wolfram cathode and cooper anode of a nozzle shape by high-frequency ignition equipment. By heating and ionization of plasma gas supplied into electric arc a high-energy plasma arc is formed by which the fed material in the form of powder is molten down, drifted, and impinges onto the machining component at high velocity. Figure 5.1 shows the basic plasma spraying equipment composing of the energy generators, of the regulation and control unit, of the plasma torch, of the cooling system, of the gas equipment, of the powder feeder, of the spraying chamber, of the exhaust system, of the shifting device for regulation of the plasma torch and of the sprayed component, and of the respective interties and hoses.

Two current generators are connected two electric network of 380 V—rectifiers with input of 2×45 kW connected into the series. The outlet of the rectifiers is connected by plasma torch with input of 45 kW, i.e. positive pole with anode a negative pole with cathode of the torch (bold connecting lines in Fig. 5.1). Regulation and control box represents a connecting element between plasma torch, rectifiers, hydraulic equipment, gas equipment and powder feeder. The regulation box automatically controls and regulates the entire spraying process. High-performance cooler in the closed circulatory system (interrupted lines in Fig. 5.1) intensively cools the plasma torch. Plasma gases of the argon or nitrogen cylinders are fed into regulation box with a single hose feeding them after being mixed in respective proportion into the plasma torch (thin lines in Fig. 5.1). Discharge pressures of gases inside the cylinders are adjusted by means of respective reduction valves with argon (or nitrogen) gas being preheated in electric preheater prior to admission into valve. Certain amount of argon of the main

© The Author(s) 2017
J. Ružbarský and A. Panda, *Plasma and Thermal Spraying*,
SpringerBriefs in Applied Sciences and Technology,
DOI 10.1007/978-3-319-46273-8_5

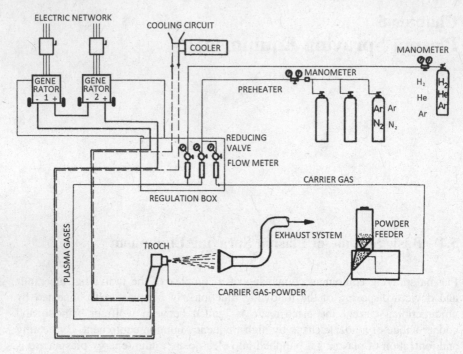

Fig. 5.1 Basic scheme of plasma spraying equipment

conducting diverted from the regulation box is supplied into the powder feeder and represents carrier gas for powder transport into the plasma torch. Powder fed into plasma jet is molten down and due to acquired kinetic energy impinges onto the sprayed component in the spraying chamber. High-performance exhaust system connected to the chamber assures exhaustion of harmful emissions and dust coming out of the working environment. Inevitable is also shifting device for movement of the plasma torch and a workpiece with respective control unit. The individual parts of device are mounted and interconnected in optimal way to provide optimal effectiveness of plasma spraying equipment with minimal energy, pressure, and spatial loses.

5.2 Plasma Torches

A plasma torch represents one of the most significant working devices of the entire system. High-energy plasma in the shape of plasma jet leaves the plasma torch nozzle (anode) at high velocity. Both electrodes are separated by the insulation intermediate part. The water under pressure cooling the torch enters the cathode part and leaves the anode part through insulation pipes or channels in insulation spacer. The torch composition must perfectly centre the cathode against the anode. Some torches are constructed to shape the electrodes and to adjust the cathodes. The plasma gas can penetrate into the space axially—through the openings inside the insulation intermediate part or tangentially—in the form of a swirl. The nozzle serving as the anode is exposed to substantial heat and must draw off several kW of input per each centimetre of a length. The nozzle is made of extremely pure copper with good conductivity and is in general directly cooled with water. The nozzle profile inside its cylindrical and conical part into which the cathode leads is determined empirically by each manufacturer.

The cathodes are composed of wolfram enriched with 2 % of ThO_2. Thorium in wolfram decreases especially the potential of electron emission and wearing of the cathode caused by impurities in plasma gases. The experience proves inevitability of massive tip. Diverse shapes of tips (thin, sharp, blunt or spherical) depend on current stress and applied plasma gases.

The most suitable gas for plasma torch is pure argon with low ionizing potential and thus electric arc can be easily ignited. The arc voltage is low since argon disposes of low ionizing voltage. Unlike argon, nitrogen and hydrogen as diatomic gases are difficult to ionize due to inevitable prior dissociation. Their arc voltage as well as ignition voltage is considerably higher contrary to argon. The advantage of the two gases rests in ability to take on extensive heat quantity.

Well-known manufacturers of plasma torches are the companies of Plasmatechnik (Switzerland) and of Metco (USA). Figure 5.2 shows the torch of 7 MB or of 9 MB by the company of Metco.

It utilizes diverse combinations of plasma gases (Ar, N_2, Ar + H_2, N_2 + H_2, Ar + He). The cooling system effects assure operation with performance of 80 kW. Through high thermic efficiency and change of electric input the enthalpy of $22.7.10^6$ J kg^{-1} can be achieved when working with pure argon or with the mixture of Ar + N_2. By structural design of the nozzle shape the exit velocity of plasma jet higher than 3000 m s^{-1} is reached. Powder particles fed by an injector placed in front of the nozzle are accelerated by the plasma jet to the velocity of 610 m s^{-1}. A switch of powder feeding is fixed onto the arm of the torch along with emergency switch that stops the entire plasma equipment in case of its failure. In dependence on used gas the appropriate nozzles mutually exchangeable can be utilized. The torch of type 7 MB-M with different external design serves for the needs of automated operation of the torch machinery by the company of Metco.

Plasma torch PH 7 S, VÚZ (Welding Research Institute) (Fig. 5.3) disposes of other equipment utilizing the non-transferred arc stabilized by tangential feed of mixture of nitrogen and hydrogen plasma gases.

Inconsiderable amount of argon is used for protection of the cathode. The torch structure is characterized by more complicated system in comparison with Fig. 5.2. The cathode 1 of thoriated wolfram adjusted in a copper holder 2 reaches the nozzle

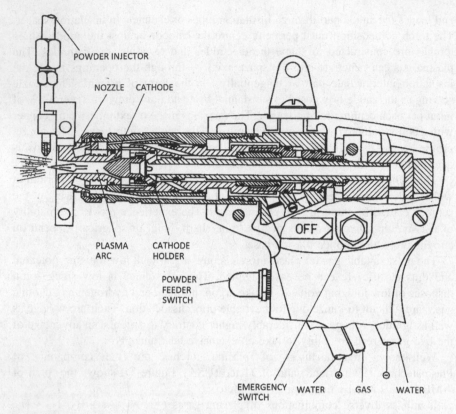

POWDER INJECTOR

NOZZLE   CATHODE

PLASMA   CATHODE
ARC   HOLDER

POWDER
FEEDER
SWITCH

OFF

EMERGENCY   WATER   GAS   WATER
SWITCH
⊕   ⊖

Fig. 5.2 Section of plasma torch MB–H by the company of Metco

I-6. The other nozzle *II-4* is separated from the preceding one by the insulation plate 7. The cathode inlet can be adjusted by loosening of a backnut 3 which may be centred by three screws 14 when the nut 5 is loosened. The cooling water is delivered by the hose in the ending of which the current supply is connected to the nozzle *II* from positive pole of source. The water is drained by the hose 12 in the ending of which the current supply is connected to the cathode from the negative pole of source. Special feeders connect the nozzle *I* and ignition electrode 15. During the torch start by condenser battery discharge into the torch between the cathode 1 and ignition electrode 15 of the first nozzle the gas is ionized—starting argon delivered by separate feeder to the area between the cathode and nozzle *I*. Thus the arc between both electrodes is ignited. Consequently, the starting argon blows the arc into the nozzle *II* by means of which the arc between the cathode and nozzle *II* is formed. At the same time the ionizer the nozzle *I* feeder are disconnected, starting argon is delivered and plasma gas supplied to the area between both nozzles is ionized. Plasma jet entering from the nozzle II melts down and carries

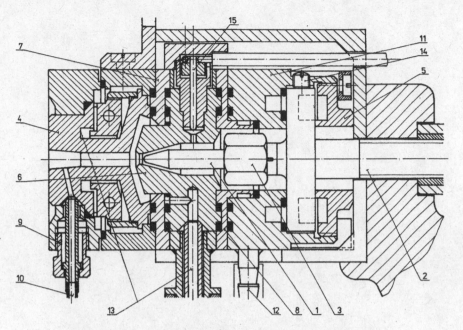

Fig. 5.3 Section of plasma torch PH 7 S, VÚZ (Welding Research Institute)

away the powder fed into the nozzle II by the carrier gas (nitrogen). During operation of the plasma torch part of argon is used for cooling and protection of the cathode. The torch with input of 25 kW with maximal voltage of 20 A requires the flow rate of 2 l min^{-1} of argon, 23 l min^{-1} of nitrogen, 8 l min^{-1} of hydrogen, and 2.5 l min^{-1} of nitrogen as of carrier gas. Consumption of cooling water with pressure ranging from 0.4 up to 0.5 MPa is of 8 even 10 l min^{-1}.

Chapter 6
Thermal Spraying

Thermal spraying allows application of the coatings resistant to abrasion and corrosion with low thermal and electric conductivity, etc. It is widely applied as a method of component renovation or in production of new film coated components with required properties of the exposed locations.

In fact, thermal spraying is application of films onto the surface of components by high thermal current containing the molten down particles or molten metal.

The influence of high velocity of solidification of the deformed particles and of their progressive accumulation results, inevitably, in the occurrence of microscopic pores, cavities which get filled by gas during formation of the film in the atmosphere. In case of metals the sprayed surface contains certain amount of oxides. If in spraying a continual diffusion coating is not formed the key problem rests in adhesion of the sprayed film onto the basis. Therefore the effective utilization of the sprayed coating requires the accurate selection of the spraying method, of parameter optimization, and knowledge of kinetics and of mechanism of film formation.

In general, thermal spraying includes three phases of a process:

1. pretreatment of basic material surface, i.e. cleaning and roughening,
2. spraying of the film,
3. machining of the sprayed film, if necessary.

According to the applied thermal energy the following variations of thermal spraying can be distinguished:

(a) gas-arc spraying in which electric arc heat is used,
(b) gas-flame spraying in which the heat generated by burning of the gas in oxygen jet or detonation wave with igniting of mixture of acetylene and oxygen is used.

© The Author(s) 2017
J. Ružbarský and A. Panda, *Plasma and Thermal Spraying*,
SpringerBriefs in Applied Sciences and Technology,
DOI 10.1007/978-3-319-46273-8_6

The spraying classification according to auxiliary material shape includes the following:

(a) wire,
(b) rod,
(c) powder.

6.1   General Characteristics of Thermal Spraying

With respect to other methods the following advantages can be ranked among general characteristics:

- Spraying causes formation of films on material with diverse properties which is not possible, e.g. spraying of ceramics, polyamides, etc.,
- In spraying the thermal influence of basic material is not that substantial during possible formation of unstable and fragile structures,
- Geometric shape of the component is not crucial, e.g. components with considerable dimensions are good to be sprayed as other methods of film application (galvanic, diffusion) cannot be taken into consideration. Only specific areas of the components can be sprayed. Due to the aforementioned the spraying appears to be highly efficient method,
- Contrary to other methods (e.g. galvanization) the thermal spraying along with welding allow application of films with thickness of several millimetres,
- Thermal spraying equipment is mostly mobile, e.g. spraying can be realized by the machine tool used for treatment of the surface, for film application, and for machining. Inconsiderable treatment is required (exhausting of waste products, protection against particle and dust spraying),
- Diverse metals can be sprayed as well as oxide ceramics, carbides, nitrides while the films with specific properties are acquired when the materials of the film are combined,
- Technological process of spraying is not arduous yet it is highly productive and effective,

Disadvantages of thermal spraying include the following:

- Thermal spraying requires assurance of good adhesion of the film, thorough abrasive pretreatment of the surface by application of crushed material (corundum, steel) and cleanness which is rather arduous in case of big components, In spraying of special coatings the reflected particles may form in combination with the air different harmful waste products damaging health,
- Due to the influence of their structure the sprayed films are not suitable for tensile stress.

6.2   Theoretical Bases of Thermal Spray Applications

6.2.1   Adhesion of Thermal Sprayed Particles to the Basic Surface and Cohesion of Film Particles

Adhesion is a complex of surface forces by which the particles of diverse materials are reciprocally attracted. Cohesion is a complex of forces by which the particles of the identical material are attracted.

Adhesion is assured by the action of forces

- in mechanical bonding,
- in physical interaction of Van der Waals type,
- in chemical interaction in separated points between basic material and coating which in several cases with metals causes local diffusion and pseudometallurgical bonds with the molten down surface of basic material by impinged particles.

Forces of the first and of the second type are inconsiderable and in electric arc spraying they reach the values measured according to DIN 50,160, i.e. of 3.5 MPa in steels. The interaction between coating material and basis follows three gradual phases in application of every elementary particle:

1. Formation of contacts, physical approximation of atoms and interaction of electron envelopes between the anchored particle and the basis. It is the first condition for further strong chemical bonds.
2. Activation and chemical interaction of atoms forming the aforementioned strong bonds due to effect of interaction of the electron envelopes of atoms and in case of atoms causing the collectivization of valence electrons in formation of new joint lattice.
3. Voluminous process with local diffusion and pseuometallurigical bonds.

Thus the adhesion of the sprayed particles depends on the following:

- temperature of particles and of basic material, i.e. temperature of contact and time of interaction,
- velocity of particles,
- state of basic material surface and reciprocal solubility of the impinging particles and the basic material.

6.2.1.1   Influence of Temperature of Particles and of Basic Material

The impingement of the molten and dispersed and transformed particle is followed by the physical contact with the boundary phase and majority of particles is ready for chemical interaction. If the number of atoms reacting during time t and situated in the proximity of the surface of each phase is denoted by x consequently the reaction velocity as per the Arrhenius equation is as follows

$$\frac{dx}{dt} = (N_o - x) \cdot v \cdot \exp\left(-\frac{E_a}{K \cdot T}\right) \exp\left(\frac{S}{K}\right) \tag{6.1}$$

with

N_o number of atoms on the surface of particle and of basic material,
v frequency of atoms,
E_a activation energy of bond formation,
S activation entropy,
K Boltzmann constant,
T absolute temperature of a contact

   For metals with lattice type of I,F or with hexagonally constrained lattice with coordination number 12 it is assumed that $\exp\left(\frac{S}{K}\right) = 1$. Consequently, the last equation is determined by the conditions of actuation of basis surface atoms and after the integration the equation with time $t = 0$ is the number of atoms on the boundary of both materials in physical contact of $x = N_o$; with time of $t = t_a$ the number of atoms having entered the reaction is of $x = N$ and the equation is as follows

$$t_a = \frac{1}{v} \ln \frac{N_o}{N_o - N} \exp \frac{E_a}{K \cdot T} \tag{6.2}$$

with

t_a time of actuation energy process

   According to the equation the relation in case of majority metals was determined $\frac{N}{N_o} = 0.6 - 0.8$. One of the main variables inevitable for determination of kinetics of reaction is temperature of contact T_k of particles with basic material. For rather short interaction time (in microseconds) and small dimension of particles T_k cannot be measured therefore calculation is used for determination. Figure 6.1 presents deformation of the particle after impingement onto the basic material.

6.2.1.2   Influence of Velocity of Particles

The arrested liquid particle during impingement onto the basic material under influence of its high kinetic energy is intensively deformed and induces considerable pressure in the zone of interaction. The generated pressure can be classified as permanent pressure R_p and impulse pressure R_i. Permanent pressure acting during particle solidification (approximately from 10^{-6} up to 10^{-7} s) is 2 even 3 times longer contrary to the impulse one. Impulse pressure causes disturbance of tiny

Fig. 6.1 Deformation of the particle after impingement onto the basic material (D_x—diameter of chemical reaction)

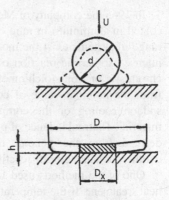

coating of turnings on the basic material surface and particle dispersion along the surface. However, permanent pressure acting during particle solidification is significant.

6.2.1.3 Voluminous Processes

Local diffusion and pseudometallurgical bonds between the particle and basic material intensively increase adhesive properties of the particle anchored into the basic material. Inevitable condition is reciprocal solubility of chemical bonds during interaction between the particle and the basic material or among the sprayed particles, which is difficult to be detected by the experiment. Its formation depends on temperature and on velocity of particles and occurs especially below the centre as well as along the particles. 10 % of bonding is of diffusion character and the rest is assured by adhesive forces.

6.2.2 Methods of Regulation of Interaction Between the Sprayed Particle Material and the Basic Material

One of the methods of basic material surface pretreatment is abrasive cleaning. The surface cleaned by the abrasive cleaning is under the influence of local plastic deformations caused by kinetic energy of impinging grains ranked among the shaped surfaces. Through the aforementioned the oxides and impurities are removed and the surface is roughened. Plastic deformation occurring in the surface crystal lattice causes formation of areas with high fluctuation of energy in case of which the energy rich zones (in the areas of failures—vacancies, dislocations, micropores) dispose of sufficiently lower energy serving for surmounting of potential barrier and for formation of chemical bonds between the particle and basic material.

In 1965 the company of Metco developed the powder composing of the nucleus coated in aluminium in ratio of 80 % of Ni and 20 % of Al. At present the entire range of systems exist the model of which is as follows: at temperature of 620 °C intense exothermic reaction occurs between Ni and Al. Nickelaluminides of stoichiometric or non-stoichiometric proportions are formed. The final phase differs from the nominal value of composition in equilibrium diagram of Ni–Al under sudden cooling of the composite particle during impingement onto the basic material. Thermal balance of physical and chemical changes of such reaction can be derived from the general relation of Joule´s law of thermodynamics. The exothermic character of reactions causes temperature increase to 3000 °C.

One of the methods used for the change of structure (of spraying) of film is its heat treatment. If the temperature of diffusion heating is increased up to proximity of alloy solid the grain boundaries occur and thus diffusion flow increases richly on the boundary of grains and environment. Continual diffusion coating forms between the film and the basic material, and porosity and from the tribological point of view unsuitable lamellar structure of the initially sprayed coating are eliminated.

The spray coating is heated by a flame (origin of the term of flame-powder welding) or in the furnace through setting-up with regular atmosphere.

Perspective is also heat treatment of coatings by laser. The temperature in heat treatment, however, influences chemical composition of powders and thus powder setting-up temperature should be lower contrary to the melting temperature of basic material. For the technology of heat treatment of spray coating the alloys of powder on the basis of NiBSi and of CoBSi usually alloyed by Cr are suitable.

6.3 Technology of Thermal Spraying

If the spraying is carried out with the auxiliary material in the shape of wire so the particle drags the compressed airstream out of the slag bath, i.e. the size of grains depends on spraying parameters. In spraying with auxiliary material in the form of powder the size of particles is determined by their grain composition. To achieve more preferable transmission of heat during the flow in the plasma jet (flame) advantageous is to dispose of particles of spherical shape which are not accelerated by the sprayed compressed airstream. The particles in dependence on spraying technology possess different temperature and size.

During impingement onto the basic material the molten particle is intensively deformed under influence of its high kinetic energy. In the particle during impingement the elastic deformation occurs and only the action of impulse pressure deforms it on the basic material surface. Each particle is subjected to thermo-mechanical process (heating of the particle, reaction with gases at higher temperatures, impingement, dispersion, and solidification). The formation scheme is shown in Fig. 6.2.

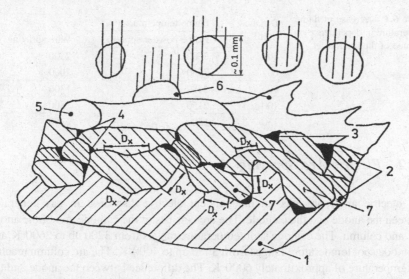

Fig. 6.2 Formation scheme of the sprayed coating. *1* basic material, *2* sprayed coating, *3* oxide, *4* pore, *5* unmolten particles, *6* particle deformed during impingement, p_x—spot of chemical interaction, *7*-mechanically wedged particle

6.3.1 Flame Spraying

In combustion of mixture of oxygen—combustible gas the flame serves as the heat source. For high temperature of even 3150 °C acetylene is used as the combustible gas. In acetylene combustion in the environment of oxygen the following chemical reaction occurs

$$C_2H_2 + 2\frac{1}{2}O_2 = 2CO_2 + H_2O + 1259.85 \quad (k\,J\,mol^{-1})$$

In wire spraying dissipation and acceleration of particles requires employment of compressed air disposing of oxidation effect upon the flame and considerably decreasing its temperature contrary to flame and powder spraying in which application of compressed air is rare. Table 6.1 presents decrease of flame temperature in dependence on distance of the torch inlet for wire and powder spraying. The respective decrease results in limitations of employment of refractory materials in flame spraying. It must be taken into consideration that in wire spraying the particle occurs at the beginning of melting and in powder spraying the same is heated during the flowing through the flame.

Table 6.1 Decrease of flame temperature in dependence on distance of the torch inlet

Distance (mm)	Flame temperature (°C)	
	Flame-powder spraying	Wire spraying
40	2600	2500
60	2550	2000
80	2480	1300
100	2150	800
150	1500	350

6.3.2   Electric Arc Spraying

The electric arc represents an independent electric discharge in gas generated between the anode and the cathode. The arc consists of the cathode and of the anode spot and column. The cathode spot temperature ranges from 3200 up to 3600 K and the anode spot temperature ranges from 3600 up to 4000 K. The arc column reaches the temperature of approximately 6000 K. The difference between the anode and the cathode temperatures rests in the fact that the cathode cools by both transmission of heat and radiation as well as by work function of electrons. The anode temperature depends on the height of the arc flowing current. In case of electric arc the principle of minimum by Stetecka is applicable according to which the electric arc is adjusted to achieve burning with minimum of electric energy with the relation as follows

$$\frac{dU}{dT} = 0 \qquad (6.3)$$

with

U voltage,
T temperature

Electric arc possesses the characteristics, i.e. dependence of voltage on current and the following is applicable for arc voltage

$$U = e_A + e_K + e_S \qquad (6.4)$$

with

e_A anode voltage drop,
e_K cathode voltage drop,
e_S arc column voltage drop

In electric arc spraying in the area of current density the characteristics is flat therefore the spraying sources dispose of flat characteristics with constant voltage (therefore voltage is regulated).

The principle of thermal spraying by electric arc (Fig. 6.3) is as follows: electric arc is formed between two fed wires out of which one represents the anode and another one represents the cathode.

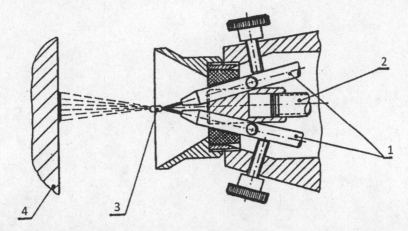

Fig. 6.3 Electric arc spraying. *1* feed of wires and current, *2* compressed air (nozzle), *3* electric arc, *4* basic material

Electric arc heat melts the auxiliary materials (wires) in the area of the anode and of the cathode spot. The compressed air flowing out of the nozzle at high velocity drags the molten particles and casts them onto the basic material. The space between the anode and the cathode is filled by the ionized gas with metal vapours which the airstream fails to drag.

Therefore the process is continual yet only with employment of unidirectional current. The feed pulleys are driven by the air-rotor or electric motor or two push-pull systems by the company of Metco.

The size of sprayed particles depends on distance between the anode and the cathode, i.e. on the arc length directly proportional to arc current. Maximal size of the particles equals to 0.1 mm. Velocity and temperature of the particles are substantially higher contrary to flame spraying. Therefore the particles sprayed by electric arc dispose of more intensive thermal impingement onto the basic material which is demonstrated by the increase of adhesion values. Such highly productive method allows spraying of diverse pseudo-alloys. Diameters of wires range from 1.6 up to 2.5 mm to prevent subsequent turbulent flowing in case of large diameter.

Disadvantages include rather intense oxidation of the sprayed metal particles especially in imperfect optimization of feed velocity of sprayed wires. Extreme overheating of the particles and their flowing within the oxidation environment result in overburning of elements by which in steel spraying the content of Si and of Mn decreases by 10 even by 15 % and the content of carbon can be decreased by 40 %.

Chapter 7
Adhesion Tests

7.1 Theoretical Analysis

The coating of the applied material is supposed to be composing of several partial coatings formed by gradual spraying of the molten particles. These individual coatings are with respect to shrinkage and to different coefficients of thermal expansibility alternately subjected to tensile and compression stress due to which crack fissures appear among the individual coatings along their boundaries and perpendicularly to the coating thickness (Fig. 7.1).

The first coating sprayed onto the basic material initially disposes of lower adhesion contrary to the adhesion of the reciprocally sprayed coatings which relates to diversity of the basic material and of the sprayed coating contrary to the uniformity of two coatings applied on each other. Therefore in case of detachment of low thickness coatings the adhesion of the basic material coating is lower contrary to the adhesion of reciprocally sprayed coatings and adhesive fracture occurs.

By means of heat effect and in extension of thickness of the sprayed coating, i.e. in application of further coatings with higher thickness more or less intensive crack growth can be observed. Consequently, in detachment of coatings from the basic material a purely cohesive fracture occurs. In case of very intensive heat effect, for instance during plasma spraying, the cracks can be healed by material melting-down and development of stress avoids the trend partially. In the first case, the one between the adhesive fracture with adhesion of σ_A and with coating thickness of x_o and the cohesive fracture with adhesion of σ_K and with the coating thickness of x_1 as shown in Fig. 7.2 with respect to rather low range of coating thickness the linear dependence can be presupposed. Subsequently the following is applicable between the coating thicknesses of x_o and x_1:

$$\sigma_x = \sigma_A - \frac{\sigma_A - \sigma_K}{x_1 - x_0}(x - x_0) \qquad (7.1)$$

© The Author(s) 2017
J. Ružbarský and A. Panda, *Plasma and Thermal Spraying*,
SpringerBriefs in Applied Sciences and Technology,
DOI 10.1007/978-3-319-46273-8_7

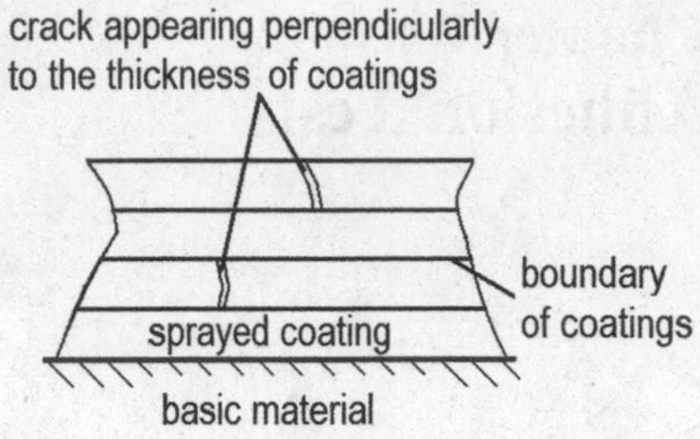

Fig. 7.1 The individual
sprayed coatings with their
boundaries and cracks
appearing perpendicularly to
the coating thickness

crack appearing perpendicularly
to the thickness of coatings

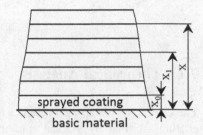

Fig. 7.2 Adhesion test of the
sprayed coating

with:

σ_x adhesion of coating with thickness x (Pa),
σ_A adhesion of coating with thickness x_0 (Pa),
σ_K adhesion of coating with thickness x_1 (Pa),
x coating thickness (m),
x_0 coating thickness (m),
x_1 coating thickness (m).

It is supposed that

$$x_0 < x_1 < x \qquad\qquad (7.2)$$

as in spraying of further coatings those are reciprocally influenced by which the
appeared cracks develop less that can be in case of significant influence expressed
by indirectly proportional dependence on the coating thickness. Consequently, the
following is generally applicable with the n exponent for the addition of dl of the
differential accrual of the crack in the distance x from the first coating:

$$dl = \frac{k}{x^n} dx \qquad\qquad (7.3)$$

with:

k constant,
n exponent.

After integration from l_o up to l and from x_0 up to x for the entire length l of the crack the following expression is applicable:

$$\int_{l_0}^{l} dl = \int_{x_0}^{x} \frac{1}{x^n} dx \tag{7.4}$$

$$l - l_0 = \frac{k}{n-1} \left(\frac{1}{x_0^{n-1}} - \frac{1}{x^{n-1}} \right) \tag{7.5}$$

and

$$l = l_0 + \frac{k}{n-1} \left(\frac{1}{x_0^{n-1}} - \frac{1}{x^{n-1}} \right) \tag{7.6}$$

with:

l_0 crack length appertaining to the coating with thickness of x_0 (m),
l crack length appertaining to the coating with thickness of x (m).

For adhesion σ_K calculated for the entire area S the following is applicable if σ_K^* is adhesion and if the ideal bonding of the areas free from defects and cracks is assured

$$\sigma_K = \sigma_K^* \frac{S - k_1 l}{S} \tag{7.7}$$

with:

k_1 constant

After substitution for l out of (7.7) the following is applicable

$$\sigma_K = \sigma_K^* \frac{S - \left[c_1 + \frac{c_2}{n-1} \left(\frac{1}{x_0^{n-1}} - \frac{1}{x^{n-1}} \right) \right]}{S} \tag{7.8}$$

with

$$c_1 = k_1 l_0 \tag{7.9}$$

$$c_2 = k_1 k \tag{7.10}$$

In general, the n exponent expresses the extent of action upon the development of cracks. In case of the least noticeable action with $n = 0$ the following is applicable

$$dl = k\,dx \tag{7.11}$$

After integration from l_0 up to l and from x_0 up to x the following is applicable for l

$$\int_{l_0}^{l} dl = \int_{x_0}^{x} k\,dx \tag{7.12}$$

$$l - l_0 = k(x - x_0) \tag{7.13}$$

and

$$l = l_0 + k(x - x_0) \tag{7.14}$$

After substitution for l into (7.7) the following is applicable

$$\sigma_K = \sigma_K^* \frac{S - [c_1 + c_2(x - x_0)]}{S} \tag{7.15}$$

In other special case the following is applicable for $n = 1$:

$$dl = \frac{k}{x}\,dx \tag{7.16}$$

After integration from l_o up to l and from x_0 p to x the following is applicable for the entire length of the crack l:

$$\int_{l_0}^{l} dl = \int_{x_0}^{x} k\,dx \tag{7.17}$$

$$l - l_0 = k\,\ln\frac{x}{x_0} \tag{7.18}$$

and

$$l = l_0 + k\,\ln\frac{x}{x_0} \tag{7.19}$$

After substituting into (7.7) the following is applicable

$$\sigma_K = \sigma_K^* \frac{S - \left(c_1 + c_2 \ln \frac{x}{x_0}\right)}{S} \tag{7.20}$$

In case of even more intensive heat influence the following is applicable for $n = 2$:

$$dl = k \frac{dx}{x^2} \tag{7.21}$$

After integration from l_0 up to l and from x_0 up to x the following is applicable

$$\int_{l_0}^{l} dl = \int_{x_0}^{x} k \frac{dx}{x^2} \tag{7.22}$$

$$l - l_0 = k \left(\frac{1}{x_0} - \frac{1}{x} \right) \tag{7.23}$$

and

$$l = l_0 + k \left(\frac{1}{x_0} - \frac{1}{x} \right) \tag{7.24}$$

After substituting into (7.7) the following is applicable

$$\sigma_K = \sigma_K^* \frac{S - \left[c_1 + c_2 \left(\frac{1}{x_0} - \frac{1}{x} \right)\right]}{S} \tag{7.25}$$

Theoretical analysis of the individual deduced functions can be performed for the diverse values n generally within the range $<0, \infty)$. For $n = 0$ achieved is the linearly decreased function with the increasing x. For $n = 1$ achieved is the logarithmically decreasing function with the increasing x and within the range for n $(1, \infty)$ achieved is hyperbolically decreasing function with the increasing x.

7.2   Experimental Conditions

Plasma and thermal spraying was performed and tested with the aim to contribute to the prolongation of service life of the tools exposed to effects of higher temperatures and to wear. Therefore sprayed were the coatings with high hardness, with resistance against wear and heat to become irreplaceable in enhancement of surface properties of the components.

7.2.1   Plasma Spraying

In the experiments used were the testing samples in case of which the coatings were applied by plasma spraying as per Tab. A.1. Those were the plasma spray coatings of the following: Mo, Al_2O_3, Al_2O_3 + 3 % TiO_2, Al_2O_3 + 13 % TiO_2, Al_2O_3 + 40 % TiO_2, $ZrSiO_4$ a ZrO_2 + 25 % MgO. These samples were prepared for testing of adhesion of plasma spray coatings.

Adhesion ranks among the basic characteristic coatings applied by plasma and thermal spraying which assesses the quality of bonding between the basic material and the coating. One of the most frequently used methods of testing focused on detection of adhesion of the coatings is the testing method according to DIN 50 160 standard or the resulting methods differing especially in dimensions of testing samples and the coating thickness. The coating is sprayed onto the degreased and roughened front area of the roll-shaped sample. After the coating has been roughened the sample is adhered to counterparts by application of adequate adhesive. The shape and dimensions of the testing samples intended for adhesion tests are shown in Fig. 7.3.

The prepared sample is subjected to tensile stress carried out by utilization of the tensile testing machine and the coating adhesion is calculated according to the following relation

$$\sigma = \frac{4 \cdot F}{\pi \cdot d^2} \tag{7.26}$$

Fig. 7.3 Shape and dimensions of testing samples intended for adhesion tests

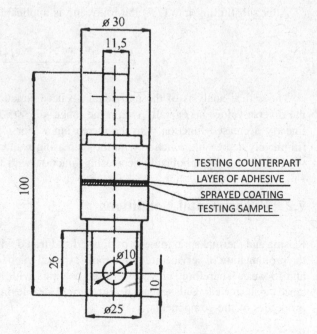

with:

σ coating adhesion (Pa),
F maximal tensile force (N),
d testing sample diameter (m)

The testing samples were gripped by the special link suspensions in order to exclude the effect of bending forces.

When the testing sample is subjected to tensile stress, the following types of disturbance can occur:

- disturbance of the interface between basic material—coating,
- disturbance of coating,
- disturbance of the interface between coating—adhesive,
- disturbance of adhesive,
- disturbance of the interface between adhesive- counterpart,
- disturbance of counterpart material.

If the adhesion test should show the assumed results, such adhesive type is then selected the strength of which is higher contrary to coating adhesion and the first two types of disturbance are taken into consideration.

During chipping and spraying the samples were screwed onto the square rod with dimensions of 10×10 mm predrilled to assure proximate gripping and intact functional areas. In the course of spraying the rod with the samples was gripped by the chuck. The angle bars with dimensions of 50×50 mm with tightly drilled holes were prepared for gripping of other samples. The cylindrical samples were inserted into the angle bar holes the balanced position of which was assured by passing of the round bar $\varnothing$ 8 mm through all holes in the samples.

The respective thickness of the coating was formed by a multiple action of the torch upon the sprayed surface in parallel lines and by its regular perpendicular descent in case of every other line on a single level so that the individual spray coating traces overlay each other and thus the even thickness of the spray coating is assured. The thickness was determined by the micrometre measurement of disc thickness when the intermediate and the functional coatings were sprayed and by comparing with the initially determined disc thickness after sand blasting. The samples were sprayed without preheating and additional cooling. The functional coatings were sprayed immediately after the intermediate coating was sprayed.

7.2.1.1 Molybdenum Coatings

Five samples were assessed (samples ranging from No. 37 up to 41, Tab. A.1). The structure is formed by the deformed lamellae formed during impingement of powder molybdenum molten in plasma. Noticeably distinguished are boundaries among the individual lamellae (Figs. 7.4 and 7.5) with lower cohesion and

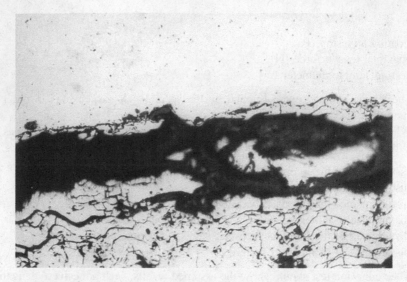

Fig. 7.4 Microstructure of the Mo coating disturbance by means of separation on the boundaries of lamellae of Mo, sample No. 6, magnification of 200×

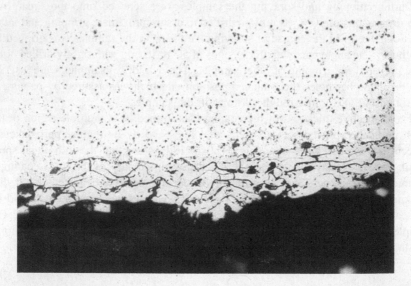

Fig. 7.5 Noticeable distinguishing of interface of lamellae of Mo, sample No. 6, magnification of 200×

disturbance occurs especially by means of separation on these boundaries. The cracks appearing crosswise the lamellae were observed as well.

A detailed view in Fig. 7.6 shows the separation on the boundaries of the lamellae and of the structures of plasma sprayed molybdenum after etching.

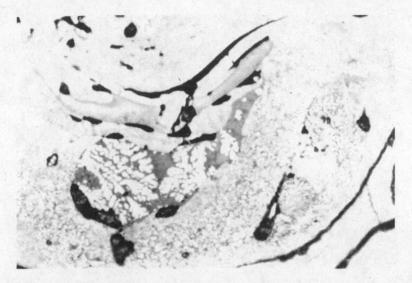

Fig. 7.6 Separation of Mo lamellae along their boundaries, the structure of plasma sprayed Mo, sample No. 5, magnification of 2000×

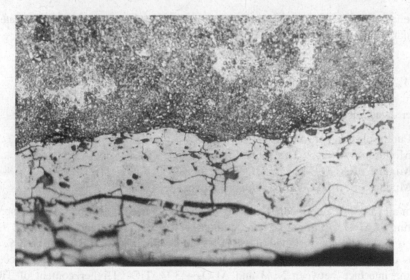

Fig. 7.7 Interface between steel and Mo coating, magnification of 2000×

The view in Fig. 7.7 documents the bond between the spray coating and the substrate in case of which visible is the fact that in plasma spraying molybdenum perfectly copies the surface of steel. However, the melt-through did not occur as the intermediate coating of Mo–Fe cannot be observed (Fig. 7.8). The Mo coating in case of sample No. 39 is substantially thicker. Insufficiently distinguished are

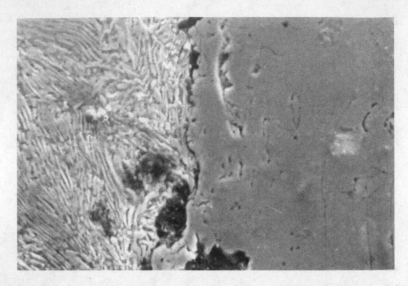

Fig. 7.8 Detail of the interface between steel and Mo, formation of the intermediate coating is not observed, sample No. 3, magnification of 1300×

lamellae of the sprayed molybdenum; the character of the bond with the substrate is similar to the one of the previous two samples.

7.2.1.2   Coatings on the Basis of Al_2O_3

After detachment tests 10 samples were subjected to examination as follows:

- Purely plasma sprayed Al_2O_3 2 samples
- Mixture of Al_2O_3 + 3 % TiO_2 3 samples
- Mixture of Al_2O_3 + 13 % TiO_2 2 samples
- Mixture of Al_2O_3 + 40 % TiO_2 3 samples

In all cases the intermediate coating of NiAl was used. Disturbance of the spray coating of Al_2O_3 occurs in fact on the interface between NiAl and Al_2O_3 (Fig. 7.9) and only locally the residues of Al_2O_3 could be documented. By adding 3 % of TiO_2 into Al_2O_3 the means of disturbance was not affected. The separation occurs on the interface between NiAl and Al_2O_3 + 3 % TiO_2. Higher content of TiO_2 increases cohesion between NiAl and the coating on the basis of Al_2O_3. In case of 40 % TiO_2 the disturbance is realized of ¾ on the interface between the intermediate coating and coating and approximately ¼ is realized in the coating. The characteristics of disturbance of the coating structure with 40 % TiO_2 is documented in Fig. 7.10 showing that the limiting factor of adhesion is not represented by cohesion among the lamellae of TiO_2 and Al_2O_3 yet by fragility of Al_2O_3. The cracks appear mainly in the lamellae of Al_2O_3.

Fig. 7.9 State of the intermediate coating of NiAl after detachment of Al_2O_3, sample No. 40, magnification of 2000×

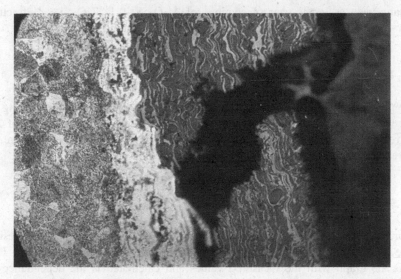

Fig. 7.10 Character of disturbance of the coating of $Al_2O_3 + 40\%$ TiO_2, sample No. 77, magnification of 200×

7.2.1.3 Coatings on the Basis of $ZrO_2 + 25\%$ MgO

Two samples were examined with different spray coating thickness and with the use of the intermediate coating of NiAl. In fact, disturbance of these coatings occurs in

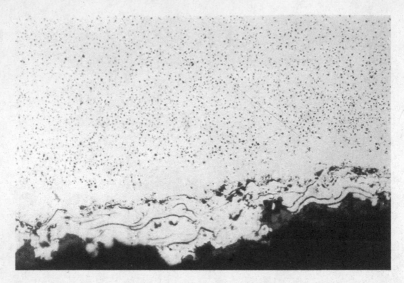

Fig. 7.11 State of the intermediate coating of NiAl after detachment of the coating of $ZrO_2 + 25$ % MgO, sample No. 51, magnification of 200×

the same way as in the case of coatings of $Al_2O_3 + 40$ % TiO_2, i.e. ¾ fracture occurs on the interface of NiAl and ¼ occurs in the coating (Fig. 7.11).

7.2.1.4 Coatings of ZrSiO₄

With the spray coating of $ZrSiO_4$ two samples with the intermediate coating of NiAl were examined. Disturbance was realized of 1/3 on the interface and of 2/3 was realized in the coating of $ZrSiO_4$. Figure 7.12 documents the structure of interface between the intermediate coating and the coating.

7.2.1.5 Measurement Results

According to the measured results of the adhesion tests the dependence of adhesion of the individual samples on the spray coating thickness was assessed. Figure 7.13 shows the graph with the measured values and calculated is the dependence of adhesion σ_K according to the relation (7.25) on the thickness x of the plasma sprayed molybdenum coating of the basic material as per STN 41 2060 standard. With the following values of $\sigma_K^* = 70$ MPa, $S = 7.065$ cm^2, $c_1 = 1.6$ cm^2, $c_2 = 0.446$ cm, $n = 2$, $x_0 = 0.01$ cm the calculated dependence according to the relation (7.25) corresponds to the measured values.

Figure 7.14 shows the plot of the measured values along with the dependence of adhesion σ_K calculated according to the relation (7.15), on the thickness x of the

Fig. 7.12 Character of the interface between the intermediate coating and the coating of $ZrSiO_4$, sample No. 45, magnification of 200×

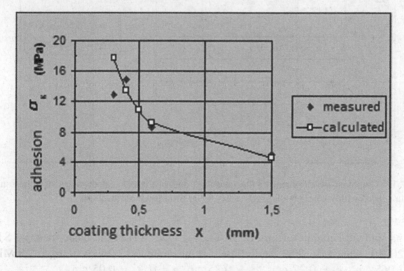

Fig. 7.13 Graphical development of the measured and of the calculated values of adhesion of plasma molybdenum spray coating on the steel basic material

plasma sprayed ceramic coating of Al_2O_3 of the basic material STN 41 2060 standard. The individual values are as follows: $\sigma_K^* = 14$ MPa, $S = 7.065$ cm^2, $c_1 = 0.22$ cm^2, $c_2 = 0.75$ cm, $n = 0$, $x_0 = 0.05$ cm.

Figure 7.15 shows the plot of the measured values and of the calculated dependence of adhesion σ_K according to the relation (7.15) on the thickness x of the

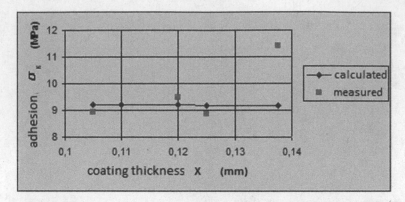

Fig. 7.14 Graphical development of the measured and of the calculated values of adhesion of plasma ceramic spray coating of Al_2O_3 on the steel basic material

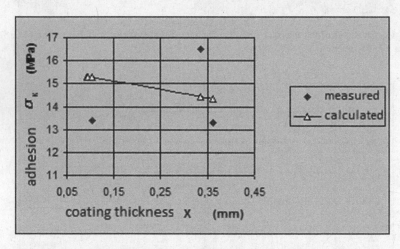

Fig. 7.15 Graphical development of the measured and of the calculated values of adhesion of plasma ceramic spray coatings of $Al_2O_3 + 13 \% TiO_2$ on the basic material

plasma sprayed ceramic coating of $Al_2O_3 + 13 \% TiO_2$ of the basic material STN 41 20260 standard. The individual values are as follows: $\sigma_K^* = 25$ MPa, $S = 7.065$ cm^2, $c_1 = 0.27$ cm^2, $c_2 = 0.75$ cm, $n = 0$, $x_0 = 0.05$ cm.

7.2.1.6 Discussion of the Achieved Results

From the point of view of cohesion the interface of the intermediate coating of Ni–Al—ceramic spray coating proved to be critical in case of double-layer spray coatings. The lowest cohesion can be observed with the coatings of Al_2O_3. By adding TiO_2 into Al_2O_3 the cohesion increases and with the content of 40 % TiO_2 it could be compared with the coatings of ZrO_2 + 25 % MgO. The most convenient cohesion on this interface is shown by $ZrSiO_4$. The best cohesion is shown by the substrate and NiAl and the intermediate coating probably due to steel surface having been treated by chipping prior to spraying. The interface between the lamellae of the intermediate coating or the Mo lamellae appeared to be the weaker point as well. The fact relates to the chemical influence of the surface of molten particles probably through oxidation. In assessment of the measured values of the adhesion to the thickness of the spray coating x it can be assumed that in case of plasma molybdenum spray coatings applicable is the derived relation (7.25) with the exponent of n = 2 and in case of plasma ceramic spray coatings applicable is relation (7.15) with the exponent of n = 0.

At the same time inevitable is to mention that the detachment tests do not take into consideration the effect of thermal shocks in case of which significant role is played by thermal stress connected with thermal expansion of the substrate and by the component of spray coating. With regards to enormous fragility of the examined ceramic plasma spray coating and to low cohesion with the intermediate coating of Ni and Al it is unlikely probable that such spray coating could withstand high thermal and dynamic stress.

7.2.2 Thermal Spraying

Basic material of thermally sprayed samples was low-carbon structural steel STN 41 1523 standard. To roughen and to acquire clean and active surface of the basic material a compact pretreatment was carried out by chipping with artificial corundum with grain composition of No.

Spraying was performed by application of the spray gun *Mogul TJ* 2 and the mixture of acetylene and oxygen was used as combustible gas. The particles were sprayed by the compressed air. The molybdenum wire with ø of 3.2 mm was used as spray coating. The chemical composition of wire was represented by 99.9 % of Mo, impurities of Al, Si, Cr, Fe, and traces of Cu and of Mg.

Tested were also the samples with molybdenum sprayed onto the basic material STN 41 1523 standard without heat treatment. With the lowest thickness of the coating the adhesive-cohesive fracture occurred and in case of higher thickness the fracture was of cohesive character. The measured values are presented in Tab. A.2 in case of which the adhesive-cohesive fracture occurred up to the thickness of 0.4 mm of the sprayed coating.

Moreover, tested was the adhesion of the sprayed molybdenum coating on the steel basic material STN 41 1523 standard after chemical and heat treatment, i.e. after two-stage nitriding of the sprayed samples.

- first temperature stage of 520 °C, heating lasting for 8 h in the furnace with ammonia atmosphere,

- second temperature stage of 560 °C, heating lasting for 8 h on the furnace with ammonia atmosphere,

Reaching and decreasing of the nitriding temperature was gradual. The measured values are presented in Tab. A.3 in case of which the adhesive-cohesive fracture occurred up to the thickness of 0.55 mm.

In regard to higher adhesive power resulting from higher value of strength of basic material surface the adhesive-cohesive fracture as well as the dependence according to the relation (4.3) was applicable in case of higher thickness contrary to previous case, i.e. up to the thickness of 0.55 mm.

Figure 7.16 shows the microstructure of the molybdenum spray coating on the basic material pretreated by chipping with artificial corundum. Figure 7.17 shows the microstructure of molybdenum sprayed coating with characteristic cracks in the spray coating layer.

Figure 7.18 shows the fracture area of cohesive fracture of the initial samples monitored by the scanning electron microscope *JMS—U3*. Figure 7.19 shows

Fig. 7.16 Microstructure of the molybdenum coating in the basic material pretreated by chipping with artificial corundum, magnification of 100×

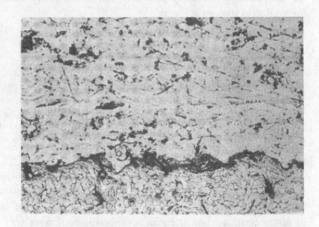

Fig. 7.17 Microstructure of the molybdenum spray coating with distinctive cracks in the spray coating layer, magnification of 500×

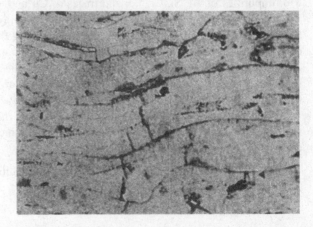

Fig. 7.18 General view of
the part of the fracture area of
cohesive fracture of the
molybdenum coating in the
basic state, magnification of
200×

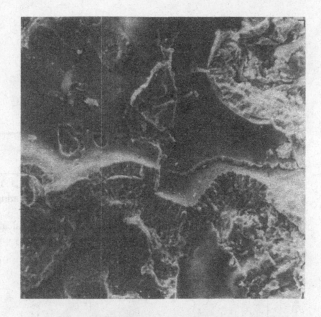

Fig. 7.19 Detail of the
fracture area of the
molybdenum coating with
distinctive orientation of
crystals in the solidified
coating, magnification of
2000×

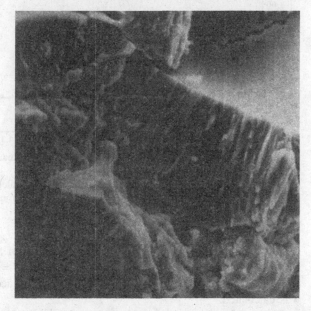

a detail of fracture area of the molybdenum spray coating with distinctive orientation of crystals in the solidified coating.

The adhesion test was followed by assessment of dependence of adhesion on the
coating thickness. Figure 7.20 shows the graph with the measured and according to
the relation (7.20) calculated dependence of adhesion σ_K on thickness of thermally

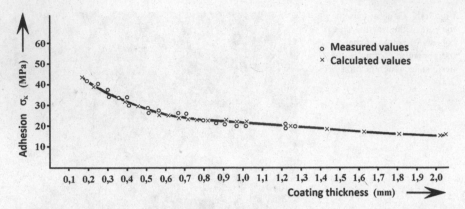

Fig. 7.20 Graphical development of the measured and of the calculated values of the adhesion of the molybdenum coatings on the steel basic material in thermally untreated state in dependence on the thickness of the coating

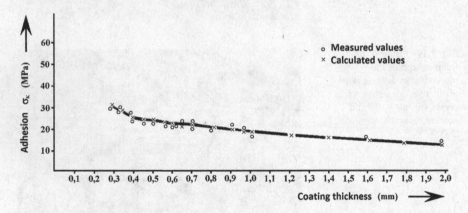

Fig. 7.21 Graphical development of the measured and of the calculated values of adhesion of the molybdenum spray coatings on the steel basic material after nitriding in dependence on the thickness of the coating

sprayed molybdenum coating x on the steel surface of STN 41 1523 standard not subjected to heat treatment. With $\sigma_K^* = 80$ MPa, $S = 7.065$ cm^2, $c_1 = 0.8$ cm^2, $c_2 = 0.537$ cm, n = 1, $x_0 = 0.04$ cm the dependence according to the relation 4.3 was calculated.

Figure 7.21 shows the graphical plot of the dependence of adhesion σ_K, calculated according to the relation (7.20), on the thickness of the sprayed coating x on the steel basic material of STN41 1523 standard after nitriding. The values identical with the ones of the previous case were achieved, i.e. $\sigma_K^* = 80$ MPa, $S = 7.065$ cm^2, $c_1 = 0.8$ cm^2, $c_2 = 0.537$ cm, n = 1, $x_0 = 0.04$ cm.

7.2.2.1 Discussion of the Achieved Results

During adhesion tests of the samples by thermal spraying of molybdenum on the thermally untreated basic material after nitriding the higher strength of the basic material surface could be observed in the second case and the adhesive-cohesive fracture along with dependence (7.20) were applicable to higher thickness contrary to the first case with the thickness of up to 0.55 mm.

So the values of adhesion increased from range of 26 up to 32 MPa in the first case with lower thickness of 0.4 mm to higher values of adhesion within the range from 25 up to 40 MPa with higher thickness of up to 0.55 mm of the sprayed coating. If compared with the dependence of the adhesion of plasma and thermal spraying the range of the measured values is approximately the same, however, in case of the plasma spray coatings more rapid adhesion decrease occurs with lower thickness of the sprayed coating.

Chapter 8
Thermal Fatigue Tests

One of the spheres of application of ceramic spray coatings are the coatings stressed by thermal shocks referred to as thermal fatigue of the coating.

8.1 Theoretical Analysis

The multiple thermal shocks, for instance, caused by sudden material heating with the liquid metal and by consequent metal cooling are referred to as thermal fatigue of the material. If: the initial temperature of the material surface is T_0, the continuous temperature is T, the heated area is S, the coefficient of the transmission of heat k, the weight of surface coating of material m, and specific heat of material c. Consequently, time differential dt passes to the material heat differential dQ:

$$dQ = k \cdot S \cdot (T_1 - T)dt = m \cdot c \, dt \tag{8.1}$$

Modification and integration of the Eq. (8.1) the following is reached

$$\int_0^t \frac{k \cdot S}{m \cdot c} dt = - \int_{T_0}^T \frac{d \cdot (T_1 - T)}{T_1 - T} \tag{8.2}$$

Through further modification after integration and with neglecting of T_0 contrary to T the following is expressed

$$T = T_1 \cdot \left(1 - e^{-\frac{k \cdot S}{m \cdot c}t} \right) \tag{8.3}$$

© The Author(s) 2017
J. Ružbarský and A. Panda, *Plasma and Thermal Spraying*,
SpringerBriefs in Applied Sciences and Technology,
DOI 10.1007/978-3-319-46273-8_8

The following can be applicable for transmission of heat from the surface coating with temperature of T_1 to the material with temperature for the time interval of dt

$$-m \cdot c \, dT = k \cdot S(T - T_1)dt \qquad (8.4)$$

After modification and integration the following is achieved

$$-\int_{T_2}^{T} \frac{d \cdot (T - T_1)}{T - T_1} = \int_{0}^{t} \frac{k \cdot S}{m \cdot c} dt \qquad (8.5)$$

After further modification and neglecting of T_1 contrary to T the following is achieved Fig. 8.1

$$T = T_1 \cdot e^{-\frac{k \cdot S}{m \cdot c}t} \qquad (8.6)$$

In plain terms, the coating of the material surface is considered to be the complex. If the coating was unoccupied and if the linear coefficient of thermal expansion α was known and with the highest surface temperature T_1 and with the temperature of the coating prior to heating T_0 its elongation could be achieved according to the following

$$\delta = \alpha \, (T_1 - T_0) \qquad (8.7)$$

The coating cannot be elongated and thus the compressive stress occurs in it:

$$\sigma = \frac{\delta \cdot m \cdot E}{m - 1} \qquad (8.8)$$

with:

m Poisson's ratio,
E elastic modulus of material

If the respective values are inserted into the relations (8.7) and (8.8)

$E = 220\,000$ MPa
$\alpha = 13.10^{-6}\,°C^{-1}$
$m = 3.3$
$T_0 = 300\,°C$
$T_1 = 600\,°C$

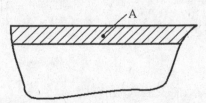

Fig. 8.1 Simplified assumption of the thermal stress of the material surface A as of the complex

The value of $\sigma = 1200$ MPa is achieved.

This compressive stress induces tensile stress in the coating below the surface coating. During the subsequent cooling of material in cooling of the surface coating having been deformed by pressure the tensile stress occurs in it and vice versa, in the coating located below the compressive stress occurs. Recurrence of such transmission as of the thermal fatigue in case of tensile stress causes the risk of crack occurrence. In order in ten thousands even in one hundred thousands the fracture actually occurs by which the service life of the material ends.

If considering the number of cycles N of material heating with thermal fatigue to be a continuous function with indirectly proportional dependence on temperature T and if every elementary increase of temperature dT means decrease of indirectly proportional cycles of service life of the material—$\frac{dN}{N}$ so thus the following is applicable

$$dT = -k\frac{dN}{N} \qquad (8.9)$$

with:

k   thermal constant (°C)

After integration from T_1 to T and from N_1 to N the following is achieved

$$T - T_1 = -k\ln N + k\ln N_1 \qquad (8.10)$$

After modification and change into common logarithm the following is achieved

$$\log N = A - K \cdot T \qquad (8.11)$$

with:

A   constant
K   constant (°C^{-1}) $K = \frac{2.3}{k}$

The relation (8.11) corresponds to the actually achieved results, for instance, in case of the mould durability in die casting according to [28] as shown in Fig. 8.2.

8.2 Experimental Conditions

In the testing of ceramic spray coating against thermal fatigue the samples according to Fig. 8.3 were used. It is a case of flat discs placed in the lower part of a chill mould into which the molten aluminium was fed. After cooling the cooled metal was taken out of the chill mould and the spray coating was air cooled.

The sample heating was dependent on the temperature of overheating of AlSi alloy. The result of the experiment and resistance of the spray coating against the

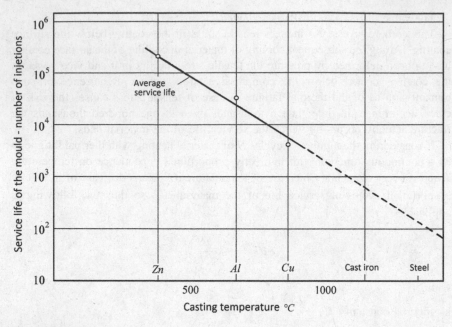

Fig. 8.2 Dependence of durability of the N cycles of the steel mould in case of die casting on the casting temperature T

Fig. 8.3 Testing sample placed in the chill mould and intendend for the thermal shock tests

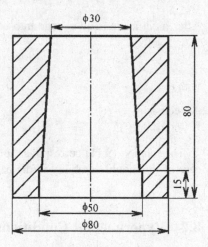

thermal fatigue are assessed according to the achieved number of cycles before the ceramic spray coating is disturbed as per Tab. A.4. Checking of the external temperature of the chill mould was performed as well which ranged from 230 up to 310 °C. The most frequent damage was detachment of the spray coating edges. According to the aforementioned tests the Al_2O_3 spray coatings with higher content of TiO_2 sprayed on the backplate made of steel of the class 19 can be regarded as

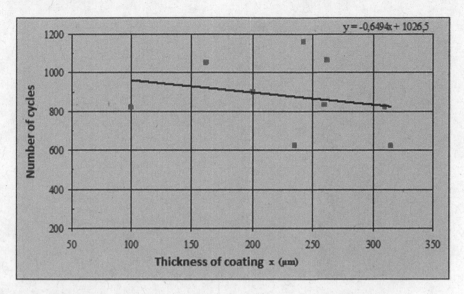

Fig. 8.4 Dependence of number of cycles on the plasma spray coating thickness

highly vulnerable to thermal fatigue. Already in case of low number of cycles the mechanical damaging and detachment of the edge in the first place occurs along with gradual damaging along the circumference of the sample. Moreover, on the basis of the aforementioned tests it could be concluded that rather significant factor in thermal fatigue is the thickness of the sprayed sample as the samples with higher thickness withstood far less cycles contrary to those with the thinner coating. The experimental dependence of the number of cycles on the thickness of the spray coating is shown in Fig. 8.4 with the regression equation.

The thermal fatigue impacts especially the components being in the contact with liquid metal in the foundry at the place of which their service life might be prolonged right by plasma spraying including the manual and mechanical ladles intended for die and chill casting, the ascending pipe for low pressure die casting, the inner part of the melting pot, etc.

Chapter 9
Roughness of Spray Coating Surface

The surface roughness is substantiated from the point of view of application of ceramic spray coatings. In the sphere of usage of ceramic spray coatings of Al_2O_3 as of the coating resistant to wear the roughness is substantially significant after mechanic machining—grinding. In this case it is inevitable to mention that the ceramic coatings on the basis of Al_2O_3 and applied by means of plasma can be machined solely by grinding. In grinding intensive cooling and use of a diamond grinding wheel is recommended. The results of surface roughness measurement are presented in Tab. A.5. The roughness was measured by the roughness measurement device by the company of Hobson–Taylor. The experimental dependence of roughness on the thickness of the spray coating is calculated according to the relation (9.1) as shown in Fig. 9.1 with the regression equation.

$$\delta = \delta_0 + k \cdot x \tag{9.1}$$

with:

δ roughness for the thickness of coating x,
x thickness of the coating (μm),
δ_0 roughness for x = 0,
k standard ($1/\mu m$)

© The Author(s) 2017
J. Ružbarský and A. Panda, *Plasma and Thermal Spraying*,
SpringerBriefs in Applied Sciences and Technology,
DOI 10.1007/978-3-319-46273-8_9

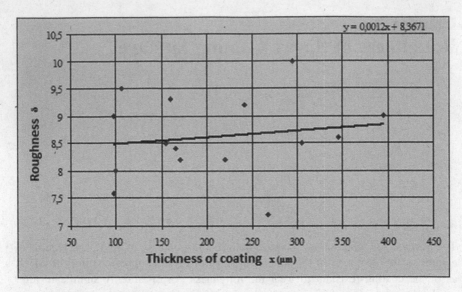

Fig. 9.1 Dependence of roughness on the plasma spray coating thickness

Chapter 10
Corrosion Tests

In case of metal structures exposed to atmospheric corrosion the films resistant to such environment are required. Increasing are the demands regarding material resistant to the environment containing salts and moisture, aggressive gases and liquids. Therefore for this purpose the corrosion tests were realized according to STN 03 8143 standard in the mist of acidified solution of sodium chloride.

The following samples were subjected to the test:

1 pc—Ø 100 chromium plated piston with the thickness of 0.01 mm Cr
1 pc—Ø 163 piston with the plasma spray coating with the thickness of 0.3–0.6 mm.

The samples were exposed to the effects of mist of 5 % NaCl solution acidified by acetic acid to pH of 3.0–3.1 at the temperature of 35 ± 2 °C in a mist chamber. The piston with plasma spray coating was conserved to the half by preservative oil OK 5 and the other half was degreased by technical petrol.

After 48 h in case of the chromium plated film the corrosion occurred on the steel basic material in the extend of 10 spots with the size of 1 mm^2. Corrosion of the sprayed material appeared after 96 h in the degreased part of the piston whereas the chromium plated film was already heavily corroded. The thickness of plasma spray coating is uneven from 0.3 up to 0.6 mm which can impact the unevenness of the corroded surface. As the basic material remained intact and disturbance affected just the plasma spray coating, the spray coating can be ground off by using the old basic material.

© The Author(s) 2017
J. Ružbarský and A. Panda, *Plasma and Thermal Spraying*,
SpringerBriefs in Applied Sciences and Technology,
DOI 10.1007/978-3-319-46273-8_10

Chapter 11
Assessment of Properties of Films on Cylindrical Testing Rods

In effort to replace the chromium plating by plasma spraying in case of the columns of metal die casting machines 5 powder types were applied onto the rods made of material 12 050 and with the dimensions of Ø 16–300 mm.

The testing rods were subjected to the following:

- tests of fatigue properties which were performed with the tensile machine with the pulsator ZD-10 PU,
- tensile tests in the sphere of plastic deformations. The tests were performed by the tensile testing machine ZD-40.

The testing conditions were selected so that they severalfold exceed standard operation loading of the material. Adhesion of the films was observed visually. In realization of the fatigue tests in order of 10^5–10^6 cycles with loading of $F_{min} = 4$ kN–$F_{max} = 70$ kN the film disturbance did not occur. The microcracks and disturbance of compactness of the films could be observed with the loadings ranging from 63 up to 80 kN. In case of the hard chromium film the microcracks occurred with the loading by the applied force of 120 kN and substantial film disturbance occurred with the loading of 127 kN.

© The Author(s) 2017
J. Ružbarský and A. Panda, *Plasma and Thermal Spraying*,
SpringerBriefs in Applied Sciences and Technology,
DOI 10.1007/978-3-319-46273-8_11

Chapter 12
Operation Tests

The draw O-rings and the press mould turrets as per Figs. 12.1 and 12.2 were produced for the operation tests. According to preselected powders the turrets and the draw O-rings were specially adjusted as Tab. A.6 shows. The aforementioned turrets and draw O-rings were subjected to operation tests.

In case of the turret with molybdenum thermal spray coating after three pressings the spray coating began to crack in the bottom part. In general, the turret was applied 80 times in pressing. On the front part in the proximity to the passage to a cone the spray coating was destroyed whereas on the conical part the spray coating remained intact, however, it contained extensive longitudinal adhering and in the calibrated part the spray coating was detached in several spots.

The O-ring of the third draw welded in the company of US Steels Košice withstood 40 pressings. After 40 pressings the O-ring disposed of the cracked welding layer of the calibration part along the entire circumference with the exception of small area of the circumference. The conic part was smooth showing no signs of wear. The O-ring of the second draw welded in the company of US Steel Košice withstood 2894 pressings. The O-ring showed in several spots, i.e. in case of 1/3 of the circumference, the detached welding layer in the calibration part. The intact welding layer in the calibration part showed extensive wear and longitudinal scratches. The conic part was smooth showing no signs of wear. Moreover, tested were the press mould turrets welded in the company of US Steel Košice which withstood approximately 50 pressings. The passage from the front part of the pressing mould turret was considerably upset. The conic part showed the longitudinal adhering identical with the one of the press mould turrets without weld deposit.

The press mould turret welded in the company of SIGMA Group a.s. Lutín withstood 60 pressings. The front of the press mould turret was smooth; the passage from the front to the conic part was partially cracked yet rather smooth. The conic part was extensively worn and cracked. Several cracked sections were completely

© The Author(s) 2017
J. Ružbarský and A. Panda, *Plasma and Thermal Spraying*,
SpringerBriefs in Applied Sciences and Technology,
DOI 10.1007/978-3-319-46273-8_12

Fig. 12.1 Press mould turret

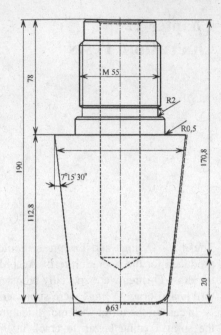

Fig. 12.2 Draw O-ring

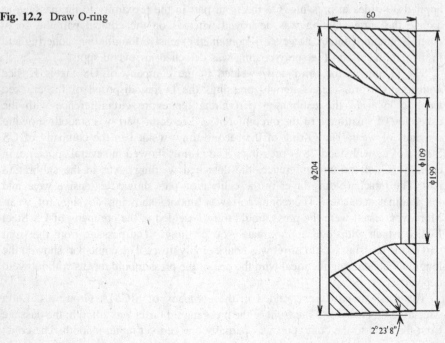

detached. The press mould turret lacked the upset neck contrary to the press mould turrets sprayed in the company of US Steel Košice.

The weld deposit samples were subjected to metallographic analysis with the use of light and REM microscopy applying the EDAX microanalysis. The preparation of samples was analogic alike in case of the spray coatings. The samples were etched in nital. Microhardness of the spray coating and of the basic material was measured with the HANEMANN microhardness meter with the loading of 100 g.

12.1 Material 12 060

The character of the interface between the weld deposit and the basic material is shown in Fig. 12.3.

As shown in the figure the plasma welding causes melting down of the basic material surface leading to local penetration of material as well as to alloying of the weld deposit with iron from the basic material. The depth of the thermally influenced zone is documented in Fig. 12.4.

The structures of the basic material occurring under the surface coating can be connected with the passed temperature cycle in case of the plasma weld deposit. For steel not being self-hardening, the structure depends on the fact if the austenitization temperature was exceeded. The area with the exceeded austenitization temperature contains the ferritic and perlitic structures formed by decomposition of austenite. In the area in case of which the temperature did not reach the austenitization temperature the martensite toughening occurs and respective structures are formed. However, contrary to basic material, in the entire influenced area the hardness decreases. The effect could be eliminated by the repeated treatment of the whole product.

Fig. 12.3 Interface between weld deposit—steel, magnification of 1000×

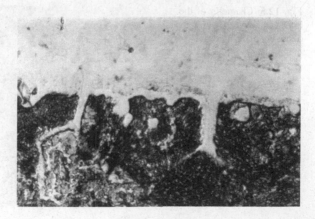

Fig. 12.4 Thermally
influenced zone,
magnification of 10×

12.2   Material 15 230

The character of the interface between weld deposit—basic material is shown in
Fig. 12.5. The size of the thermally influenced area can be assessed according to
Fig. 12.6.

.Contrary to the previous steel the basic material—steel with higher
through-hardenability in the part of the influenced zone with the exceeded
austenitization temperature during cooling the occurrence of unbalanced decomposition to martensite can be observed. In the area in case of which the temperature
did not reach the austenitization temperature the resulting structures are of ferritic
and carbide type, i.e. the structures of more intensive martensite toughening unlike
in case of basic material. On the interface between the thermally influenced zone
and the basic material the path of cracks appears (Fig. 12.7). The cracks are
structured, i.e. probably it is not the case pf splitting of a crack.

Fig. 12.5 Character of the
interface between weld
deposit—basic material,
magnification of 160×

Fig. 12.6 Thermally
influenced zone,
magnification of 10×

Fig. 12.7 Cracks appearing
on the interface between the
thermally influenced zone—
the basic material,
magnification of 160×

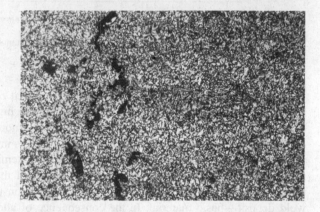

It can be assumed that in case of the aforementioned steel the toughening fragility was demonstrated and disturbance along the boundaries of grains with action of dilatation strain occurred.

12.3 Material 19 541

The resulting structure in the passage section reflects the past temperature cycle in plasma welding. The basic material is represented by tool steel possible to be hardened in the air. In the area with the exceeded austenitization temperature the martensitic transformation occurred during cooling. Martensite morphology depends on the temperature of hardening which is related to the size of austenite grain and consequently to the size and to the shape of martensitic plates. In case of press mould turrets the metallographic analysis included measurement of microhardness of the structures in the passage area. The measured values are shown in Fig. 12.8.

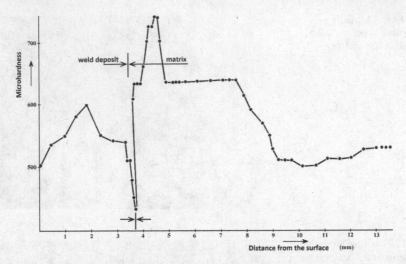

Fig. 12.8 Dependence of microhardness HVM 100 on the distance of the weld deposit external surface

If hardness of the basic material in the uninfluenced zone is compared with the hardness of the weld deposit observed can be slightly higher hardness of the weld deposit at room temperature. The situation becomes more favourable for the weld deposit at increased temperatures as the weld deposit is in fact the cobalt superalloy maintaining the strength characteristics for higher temperatures alike the basic material. From such point of view the application of the weld deposit would be reasoned. However, rather adverse situation occurs on the interface between the weld deposit—basic material. In the consequence of alloying the superalloy with iron considerable decrease of hardness on the interface can be observed. The cause could also rest in alloying of the basic material, in stabilization of austenite, and large amount of residual austenite remaining on the interface after cooling. The result is that the structure is formed on the interface possessing substantially lower mechanical properties contrary to the weld deposit and basic material. Initially, in the passage area rapid increase of the hardness can be observed (which can be related to dissolution of carbides and to alloying of austenite) and consequently the zone with constant or high hardness. Following is the passage of hardness decrease (low austenitization temperature—a few soluble carbides) to the values lower contrary to those of material in basic refined state (highly toughened martensite).

Fig. 12.9 Crack occurring in the weld deposit and in the thermally influenced zone and ending in the thermally influenced zone, magnification of 10×

Fig. 12.10 Local cracks in the thermally influenced zone and the incipient crack on the interface between weld deposit—steel, magnification of 10×

Operation tests concerning the press mould turret was followed by occurrence of the cracks in the weld deposit spreading towards the basic material (Fig. 12.9). Their length is influenced by the thickness of passage coating thermally influenced during plasma welding. The independent cracks were also observed in the thermally influenced coating (Fig. 12.10).

Chapter 13
Conclusion

In the final chapter the components taken into consideration for plasma spraying can be analysed and classified into several categories.

1. The tools for special production operating under heat are the following: press mould turret, upsetting head, and draw O-ring.

   During tests the service life comparable with the current state or maximally by 20 % higher was reached. The saving is not provable to such extent so that plasma spraying could be used under given circumstances. The tool service life terminates after its bottom part is upset which is proved by the observation and by the undesirable high thermal influence of the tool and by the absolute numbers of cycles moving on the level of 50–70 before the service life termination. Therefore considered is the press line with the exclusion of human factor and decrease of thermal influence of the tool. In case of such new production assumed is testing of the tools with plasma spray coating under the condition that the service life is terminated by the surface grinding which in plasma spray coating would means removing of the spray coating by blast cleaning. Consequently, the tool could be repeatedly sprayed.

2. The tools for special production operating under cold conditions are the following: a draw arbor, a draw O-ring, an arbor, a matrix, a header pin, a header matrix, and bottom part of the headering.

   Service life of these components terminates by grinding or by blast cleaning of certain part of the surface. It is assumed that in plasma spraying the situation improves when such part of the coating is ground or detached which can be restored after blast cleaning.

3. The components designed for die casting machines, the machines designed for plastic coating spraying machines and manipulators in case of which the sprayed coating replaces higher quality material treated thermally. The category includes the skids for die casting machines, bars designed for grinding machines, skids

© The Author(s) 2017   99
J. Ružbarský and A. Panda, *Plasma and Thermal Spraying*,
SpringerBriefs in Applied Sciences and Technology,
DOI 10.1007/978-3-319-46273-8_13

for manipulators, flanges of closures designed for die casting machines, nozzles, screw taps and extension pieces for screw taps for plastic coating spray machines. Plasma spraying was applied in case of skids for die casting machines, bars for grinding machines and nozzle for plastic coating spray machines. According to the tests the bars for grinding machines proved to be comparable with currently used bars.

4. The components for die casting machines, plastic coating spray machines, and auxiliary machines—rejects repaired by chromium plating, renovated components repaired by chromium plating that can be repaired by plasma spraying, and components for die casting machines, hydraulic cylinders, telescopic hydraulic motors in case of which pistons must be protected against liquid metal or corrosive environment. It includes the columns of die casting machines, the piston of pressing, the pistons of smoothing cylinders, the piston of closure, the pistons of hydraulic cylinders, and the telescopic hydraulic motors. The rods with Ø of 16 × 250 mm were plasma sprayed and were subjected to fatigue and tensile testing. The telescopic hydraulic motor TGG 62 was plasma sprayed and subjected to chromium plating in the mist of acidified solution of sodium chloride in the condensing chamber. Corrosion of the chromium plated coating appeared after 48 h and corrosion of the plasma spray coating could be observed after 96 h.

5. The components being in contact with the liquid metal in the foundry in case of which the plasma spray coating could prolong the service life. The category includes the following: the manual and mechanical ladles intended for die and chill casting, the ascending pipe for low pressure die casting, the inner part of the melting pot, and possibly the external surface of the gooseneck immersed in a liquid metal. Realized were tests concerning the samples with plasma spray coating by liquid aluminium casting. The samples withstood 1000 castings without coating disturbance. The components being under maintenance are taken into consideration for renovation in case of which the service life could be prolonged.

6. A complement could be represented by the components intended for chemical industry, especially thread advancing reel of diverse dimensions.

Annexes

See Tables A.1, A.2, A.3, A.4, A.5 and A.6.

Table A.1 Testing samples with the spray coating for the adhesion tests with the assessed results

Sample No.	Kind of spraying	Thickness of spraying (μm)	F (0.01 MN)	σ (MPa)	Notes
1	Al_2O_3	100–110		2.26	Cracked in the centre
2	Al_2O_3	100–110	0.63	8.917	
3	Al_2O_3	105–115	0.4	5.662	
4	Al_2O_3	115–125	0.67	9.483	
5	Al_2O_3	120–130	0.625	8.846	
6	Al_2O_3	125–150		11.4	Cracked in the centre
7	$ZrSiO_4$	220–280	0.73	10.332	
8	$ZrSiO_4$	250–300		3.75	Cracked in the centre
9	$ZrSiO_4$	300–400		2.97	Cracked in the centre
10	$ZrSiO_4$	300–430	0.95	13.446	
11	$ZrSiO_4$	300–450	0.86	12.172	
12	$ZrSiO_4$	400–500		9.2	Cracked in the centre
13	$ZrO_2 + 25~\%~MgO$	65–80		3.54	Cracked in the centre
14	$ZrO_2 + 25~\%~MgO$	65–85			
15	$ZrO_2 + 25~\%~MgO$	65–95	Fallen off during manipulation		
16	$ZrO_2 + 25~\%~MgO$	115–150	0.37	5.237	Cracked in the centre
17	$ZrO_2 + 25~\%~MgO$	115–180	0.49	6.935	
18	$ZrO_2 + 25~\%~MgO$	130–210		5.94	Cracked in the centre
19	$Al_2O_3 + 3~\%~TiO_2$	105–135		15.7	Cracked in the centre
20	$Al_2O_3 + 3~\%~TiO_2$	125–150	1.18	16.7	
21	$Al_2O_3 + 3~\%~TiO_2$	135–175	0.98	13.871	
22	$Al_2O_3 + 3~\%~TiO_2$	270–360		5.66	A–K
23	$Al_2O_3 + 3~\%~TiO_2$	275–360	1.235	17.48	
24	$Al_2O_3 + 3~\%~TiO_2$	325–480		5.76	A–K

(continued)

© The Author(s) 2017
J. Ružbarský and A. Panda, *Plasma and Thermal Spraying*,
SpringerBriefs in Applied Sciences and Technology,
DOI 10.1007/978-3-319-46273-8

Table A.1 (continued)

Sample No.	Kind of spraying	Thickness of spraying (μm)	F (0.01 MN)	σ (MPa)	Notes
25	Al_2O_3 + 13 % TiO_2	73–115		12.173	K
26	Al_2O_3 + 13 % TiO_2	80–115	0.82	11.606	
27	Al_2O_3 + 13 % TiO_2	93–115	0.945	13.375	
28	Al_2O_3 + 13 % TiO_2	230–440		5.94	A–K
29	Al_2O_3 + 13 % TiO_2	230–440	1.165	16.485	
30	Al_2O_3 + 13 % TiO_2	275–440	0.94	13.305	
31	Al_2O_3 + 40 % TiO_2	275–365		11.4	A–K
32	Al_2O_3 + 40 % TiO_2	360–440		15.853	K
33	Al_2O_3 + 40 % TiO_2	440–490	0.42	5.945	
34	Al_2O_3 + 40 % TiO_2	440–530		3.75	A–K
35	Al_2O_3 + 40 % TiO_2	420–600	0.97	13.729	
36	Al_2O_3 + 40 % TiO_2	420–600	0.945	13.375	
37	Mo	0.3 mm	0.91	12.88	
38	Mo	0.4 mm	1.05	14.86	
39	Mo	0.5 mm	1.037	19.39	
40	Mo	0.6 mm	0.615	8.704	
41	Mo	1.5 mm	0.35	4.954	

Table A.2 The measured and the calculated values of adhesion of molybdenum thermal spray coatings on the steel basic material in thermally untreated state

Sample No.	Coating thickness (μm)	Measured adhesion (MPa)	Calculated adhesion (MPa)	Notes (fracture)
1	300	31,950	31,950	A–K
2	350	29,580	29,010	A–K
3	400	26,890	26,080	A–K
4	400	27,030	26,080	A–K
5	400	27,370	26,080	A–K
6	400	24,340	26,080	A–K
7	400	24,370	26,080	A–K
8	400	25,760	26,080	A–K
9	450	24,620	25,400	K
10	450	22,360	25,400	K
11	500	22,640	24,728	K
12	500	22,940	24,728	K
13	500	25,190	24,728	K
14	550	23,920	24,168	K
15	600	21,510	23,616	K
16	600	21,650	23,616	K

(continued)

Table A.2 (continued)

Sample No.	Coating thickness (μm)	Measured adhesion (MPa)	Calculated adhesion (MPa)	Notes (fracture)
17	650	21,790	23,144	K
18	650	23,770	23,144	K
19	700	25,190	22,680	K
20	700	19,950	22,680	K
21	800	20,510	21,864	K
22	800	21,230	21,864	K
23	900	23,350	21,152	K
24	950	21,510	20,832	K
25	1000	18,400	20,504	K
26	1000	19,390	20,504	K
27	1200	–	19,656	–
28	1400	–	18,464	–
29	1600	18,580	17,648	K
30	1600	17,550	17,648	K
31	1800	–	16,936	–
32	2000	17,120	18,296	K
33	2000	17,410	16,296	K

Table A.3 The measured and the calculated values of adhesion of molybdenum thermal spray coatings on the steel basic material after chemical and thermal treatment—i.e. nitriding

Sample No.	Coating thickness (μm)	Measured adhesion (MPa)	Calculated adhesion (MPa)	Notes (fracture)
1	200	40,620	40,335	A–K
2	200	40,050	40,335	A–K
3	220	39,770	39,411	A–K
4	250	38,210	38,025	A–K
5	300	35,810	35,716	A–K
6	300	34,670	35,716	A–K
7	350	33,120	33,407	A–K
8	400	33,120	31,096	A–K
9	400	30,000	31,096	A–K
10	450	28,300	28,788	A–K
11	500	27,600	26,478	A–K
12	500	26,180	26,478	A–K
13	500	25,190	26,478	A–K
14	550	25,050	24,168	A–K
15	600	–	23,616	–
16	650	24,340	23,144	K
17	700	24,200	22,680	K

(continued)

Table A.3 (continued)

Sample No.	Coating thickness (μm)	Measured adhesion (MPa)	Calculated adhesion (MPa)	Notes (fracture)
18	700	23,070	22,680	K
19	750	22,950	22,270	K
20	800	21,750	21,864	K
21	850	20,800	21,504	K
22	900	–	21,152	–
23	950	19,810	20,832	K
24	1000	19,390	20,504	K
25	1200	19,280	19,656	K
26	1250	18,960	19,280	K
27	1400	18,820	18,464	K
28	1400	18,680	18,464	K
29	1600	–	17,648	–
30	1800	–	16,936	–
31	2000	–	16,296	–

Table A.4 Testing samples with plasma spray coating for the thermal fatigue tests

Sample No.	Basic material	Spray coating	Thickness (μm)	Number of cycles	Notes
1	19 191	$ZrSiO_4$	300 - 330	620	
2	19 312	$Al_2O_3 + 40\%\ TiO_2$	180 - 220	900	
3	19 312	Al_2O_3	95 - 105	820	
4	19 452	$Al_2O_3 + 13\%\ TiO_2$	230 - 290	832	
5	19 721	$Al_2O_3 + 40\%\ TiO_2$	240 - 285	1062	
6	19 721	Al_2O_3	105 - 115	670	
7	12 050	$Al_2O_3 + 40\%\ TiO_2$	150 - 175	1050	
8	11 500	$Al_2O_3 + 40\%\ TiO_2$	225 - 260	1160	
9	11 500	$ZrSiO_4$	270 - 350	820	
10	12 050	$Al_2O_3 + 13\%\ TiO_2$	220 - 250	620	
11	19 191	Al_2O_3	80 - 90	260	
12	42 2420	Al_2O_3	160 - 180	1 - 2	In case of these samples used as the pistons for filling chamber in die casting already during the first pressing detachment of the sprayed coating could be observed.
13	42 2420	$ZrSiO_4$	550 - 650		
14	42 2420	$ZrO_2 + 25\%\ MgO$	250 - 330		
15	42 2420	$Al_2O_3 + 13\%\ TiO_2$	440 - 540		

Table A.5 Measured values of roughness of the plasma sprayed samples

Sample No.	Basic material	Spray coating	Thickness (µm)	Roughness
1	19 191	Al_2O_3 + 40 % TiO_2	225–260	9.2
2	19 191	Al_2O_3 + 40 % TiO_2	270–320	10
3	19 191	Al_2O_3 + 13 % TiO_2	235–300	7.2
4	19 191	Al_2O_3	100–115	9.5
5	19 191	Al_2O_3 + 13 % TiO_2	290–320	8.5
6	19 191	$ZrSiO_4$	350–440	9
7	19 312	Al_2O_3 + 13 % TiO_2	180–260	8.2
8	19 312	$ZrSiO_4$	330–360	8.6
9	19 434	Al_2O_3 + 40 % TiO_2	150–180	8.4
10	19 434	Al_2O_3 + 13 % TiO_2	155–185	8.2
11	19 434	Al_2O_3	95–100	7.6
12	19 434	$ZrSiO_4$	330–360	<10
13	19 462	Al_2O_3 + 40 % TiO_2	140–180	9.3
14	19 652	Al_2O_3	95–105	8
15	19 452	$ZrSiO_4$	330–360	<10
16	19 721	Al_2O_3 + 40 % TiO_2	240–285	<10
17	12 050	Al_2O_3	95–100	9
18	12 050	$ZrSiO_4$	300–360	<10
19	11 500	Al_2O_3 + 13 % TiO_2	135–175	8.5

Table A.6 Plasma welded samples for the operation tests

Sample No.	Name of sample	Dimension prior to welding	Dimension after welding	Basic dimension	Dimension after grinding	Note
1	Draw O-ring	Ø 112.2	Ø 108.2	Ø $109.4^{+0.2}$	Ø 109.49	Weld deposit performed with powder K 40
2	Draw O-ring	Ø 112.0	Ø 109.5	Ø $109.4^{+0.2}$	Ø 111.60	Weld deposit performed with powder K 45
3	Draw O-ring	Ø 112.1	Ø 109.0	Ø $109.4^{+0.2}$	Ø 111.55	Weld deposit performed with powder K 50
4	Punching head	Ø 91.37	Ø 91.8	Ø $91.79^{-0.1}$	Ø 91.24	Weld deposit performed with powder K 55
5	Punching head	Ø 91.14	Ø 91.8	Ø $91.79^{-0.1}$	Ø 91.78	Weld deposit performed with powder K 40
6		Ø 91.8	Ø 91.79		Ø 91.75	

(continued)

Table A.6 (continued)

Sample No.	Name of sample	Dimension prior to welding	Dimension after welding	Basic dimension	Dimension after grinding	Note
	Punching head			Ø $91.79^{-0.1}$		Weld deposit performed with powder K 55
7	Punching head	Ø 91.21	Ø 92.0	Ø $91.79^{-0.1}$	Ø 91.73	Weld deposit performed with powder K 50
8	Punching head	Ø 91.04	Ø 91.6	Ø $91.79^{-0.1}$	Ø 91.74	Weld deposit performed with powder K 50
9	Punching head	Ø 91.24	Ø 91.7	Ø $91.79^{-0.1}$	Ø 91.74	Weld deposit performed with powder K 50
10	Punching head	Ø 91.44	Ø 92.0	Ø $91.79^{-0.1}$	Ø 91.75	Weld deposit performed with powder K 40
11	Punching head	Ø 91.19	Ø 91.6	Ø $91.79^{-0.1}$	Ø 91.67	Weld deposit performed with powder K 50
12	Punching head	Ø 91.30	Ø 91.7	Ø $91.79^{-0.1}$	Ø 91.72	Weld deposit performed with powder K 55
13	Punching head	Ø 91.20	Ø 91.8	Ø $91.79^{-0.1}$	Ø 91.72	Weld deposit performed with powder K 40

Bibliography

1. Ambrož, O.: Vytváření povlaků metódami žárového nástřiku a jejich využití v prvovýrobe a renovacích. In Zváranie **7**, 152 (1993)
2. Ambrož, O., Kašpar, J.: Žárové nástřiky a jejich průmyslové využití. STNL, Praha (1990)
3. Blaškovič, P., Čomaj, M.: Renovácia naváraním a žiarovým striekaním, p. 378. DT ZSVTS, Žilina (1991)
4. Čabelka, J.: Mechanická technológia, p. 1036. Vydavateľstvo SAV, Bratislava (1967)
5. *HS Technik: Ochranné povlaky* [cit. 2005-02-02]. The internet: http://www.hstechnik.sk/
6. Chasuj, A.: Technika napylenija. Moskva (1975)
7. Kováč, I., Tolnai, R., Žarnovský, J.: Vplyv bóru na zvýšenie mechanických vlastností povrchových vrstiev ocelí. In: Kvalita a spoľahlivosť strojov, pp. 75–77. SPU, Nitra (2000). ISBN 80-7137-720-1
8. Kreibich, V.: Povrchové úpravy. Edičné stredisko ČVUT, Praha (1982)
9. Laudar, J.: Liatie pod tlakom, p. 268. SVTL, Bratislava (1964)
10. Maňas, S.: Hydraulické mechanismy strojů a zařízení. ČVUT, Praha (1991)
11. Matejka, D., Benko, B.: Plazmové striekanie kovových a keramických práškov, p. 271. ALFA, Bratislava (1988)
12. Ragan, E.: Väzkosť a počiatočné napätie pri zliatinách v intervale kryštalizácie. FVT TU v Košiciach, Prešov (1997)
13. Ragan, E., Ružbarský, J.: Stanovenie parametrov v dutine formy. In: Materiálové inžinierstvo, roč. XIII, 2006, No. 21–22. ISSN 1335-0803
14. Ragan, E., Ružbarský, J.: Pôsobenie kvapalného kovu na kovové časti formy a stroja pri tlakovom liati. In: Kvalita a spoľahlivosť technických systémov, pp. 164–165. SPU Nitra, Nitra (2006). ISBN 80-8069-707-8
15. Ragan, E., Ružbarský, J., Andrejčák, I.: Vývoj a perspektívy liatia pod tlakom. In: Slévárenství, roč. LIII, No. 2–3, pp. 84–87 (2005). ISSN 0037-6825
16. Ruml, V.: Moderné úpravy kovového povrchu. Práca, Bratislava (1959)
17. Ružbarský, J., Ragan, E.: Dynamika procesu plnenia lyžice zariadenia pre dávkovanie kovu pri liati pod tlakom. In: Elektronický zborník Nové trendy v prevádzke výrobnej techniky 2003, VI. medzinárodná vedecká konferencia, 20–21 Nov 2003, pp. 355–358, FVT Prešov, Prešov (2003). ISBN 80-8073-059-8
18. Ružbarský, J., Ragan, E.: Regulácia procesu dávkovania kovu pri liati pod tlakom. In: Kvalita a spoľahlivosť technických systémov, pp. 165-168. SPU Nitra, Nitra (2005). ISBN 80-8069-518-0
19. Ružbarský, J., Ragan, E.: Trvanlivosť foriem a statných častí tlakových lejacích strojov vzhľadom na styk s tekutým kovom. In: Materiálové inžinierstvo, roč. XIII, No. 3, pp. 77–78 (2006). ISSN 1335-0803
20. Ružbarský, J.: Viskozita a počiatočné napätie zliatin v intervale kryštalizácie. In: Technologické inžinierstvo, roč. IV, No. 1, pp. 117–119 (2007)

© The Author(s) 2017
J. Ružbarský and A. Panda, *Plasma and Thermal Spraying*,
SpringerBriefs in Applied Sciences and Technology,
DOI 10.1007/978-3-319-46273-8

21. Ružbarsky, J.: Rozbor priľnavosti striekaných vrstiev na základnom materiáli. In: Výrobné inžinierstvo, roč. 2, No. 2–3, pp. 22–25 (2003)
22. Ružbarský, J.: Vývoj technológie liatia železných zliatin pod tlakom. Technologické inžinierstvo, roč. IV **1**, 115–116 (2007)
23. Stejskal, M.: Československé tlakové licí stroje. Racionalizace a automatizace tlakového lití. Sborník z konference 22–24.6.1976, Gottwaldov
24. Šmiga, S.: Automatizované pracoviská tlakového liatia CLHA 400.05. In: Slevárenství, No. 3 (1993)
25. Takáč, K.: Technológia povrchových úprav, p. 160. ALFA, Bratislava (1988)
26. Trnková, L.: Zborník vedeckých prác MtF STU, zv. 9, p. 105. Vydavateľstvo STU, Bratislava (2000)
27. Trojánek a kol, F.: Příručka pro povrchové úpravy IV. SNTL, Praha (1964)
28. Uzík, L.: Sortiment a vlastnosti práškových prídavných materiálov pre plameňopráškové naváranie. Zaškolenie na plameňovo – práškové nanášanie, p. 17. Dom techniky, ČSVTS, Žilina (1987)
29. Valecký a kol, J.: Lití kovu pod tlakem. SNTL, Praha (1963)
30. Vasilko, K., Kmec, J.: Delenie matériálu, p. 232. Data Press, Prešov (2003)
31. Vasilko, K., Kožuro, L., Ružbarský, J.: Mechanizmus elektromagnetického naváranie povrchovej vrstvy na súčiastky. In: Zborník zo 6. medzinárodnej vedeckej konferencii Transfer 2004, pp. 544–549. GC-TECH, Ing. Peter Gerši, Trenčín (2004). ISBN 80-8075-030-0
32. Prospekt Kovové prášky, p. 2. Výskumný ústav zváračský, Bratislava (2000)

Printed in the United States
By Bookmasters